# mobile media learning

*amazing uses of mobile devices for learning*

**EDITED BY:**
Seann Dikkers
John Martin
Bob Coulter

**mobile media learning:**
*amazing uses of mobile devices for learning*

*Design & composition by Szu Yuan Scott Chen*
*(www.scottchendesign.com)*

# foreword/preface

## by Eric Klopfer

I was sitting in a meeting recently listening to a presenter from a major media company talk about mobile device usage by kids. The crowd quickly reacted to the "dated" data on apps and devices, hankering for more up to date information. The data was only a year old, but in the rapidly moving space of mobile media, that data is already stale. While some may lament the rapid pace of change that seems to leave devices themselves to become quickly dated, I see the vast opportunities that this rapid change has created for both developers and users of mobile applications. We've come a long way in the last ten years, creating a norm that requires justification for developing applications for anything other than mobile platforms. That is a dramatic shift from the questions we faced a decade ago, which forced us to justify our pursuit of learning applications on what we now recognize as primitive mobile devices.

Just what is "mobile" in this context? I recently passed by a wheeled cart with half a dozen desktop computers complete with giant CRT monitors. The cart had a big sign labeling it as a "Mobile Computer Lab". This is not mobile. Laptops aren't mobile in this context either. Mobility requires the ability to casually use a device on the go — without sitting down. While this may seem arbitrary it is important in that it influences the way we use these devices. They aren't only for long focused sessions, but can also be used for bite–sized interactions taking place for mere seconds in the context of some other related (or unrelated) activity. This is the first of several defining factors that Kurt Squire and I (Klopfer

& Squire 2007) outlined as being unique affordances of mobile devices for learning. The relevance of these five affordances has only increased over time.

- *portability* — can take the computer to different sites and move around within a location
- *social interactivity* — can exchange data and collaborate with other people face to face
- *context sensitivity* — can gather data unique to the current location, environment, and time, including both real and simulated data
- *connectivity* — can connect handhelds to data collection devices, other handhelds, and to a common network that creates a true shared environment
- *individuality* — can provide unique scaffolding that is customized to the individual's path of investigation.

The devices we had at hand at the time were stylus driven Palms and brand new color Pocket PCs that allowed us to use wireless connectivity or GPS (but not both simultaneously). We were convinced that these devices would some day be ubiquitous enough where the benefits of these devices could be realized inside the classroom and out. Teachers liked the simplicity and students liked the personalized experience. Surely the day would come when every kid would have a Palm or Pocket PC in their hand.

While we may not have bet on the exact right horse, that day of ubiquitous access to mobile devices is almost upon us — if only schools wouldn't ban them. Yet, the scale of success of mobile platforms, as well as the kinds of applications and modes of interaction, are well beyond what I could have expected. This came from both developers and users embracing mobile for its affordances, rather than trying to scale down full sized apps to a smaller screen. I remember when satirical newspaper, The Onion, released their Palm app with the slogan "The Onion just got smaller and harder to read." But that is indeed what happened, and still continued to happen to some degree.

Early developers of games on mobile platforms struggled to figure out how they would get their games to work on these devices without a D– Pad (four or eight way controller found on many mobile and fixed gaming consoles). Even many of the early iPhone games tried to figure out ways to make use of an onscreen D– Pad. But then games like Angry Birds and Cut the Rope, which were uniquely suited to the touch screen started to emerge. These were games that used the mobile interface to their advantage and helped developers and users realize that we should design for these devices, not around them. Now the tables have been turned and makers of gaming consoles are trying to figure out how to incorporate some of the mobile interaction style into their own platforms. It turns out that in many ways mobile devices are easier to use for many tasks that take into account personal information, location, and more natural interactions like touch and voice. This allows the user/learner to focus on what they are doing instead of how they are doing it.

The experiences and activities in this book are not ones that have been designed to work around the shortcomings of mobile devices, but rather are

designed to take advantage of what
is special and unique about them. They
are also leading the field into new directions that show what is powerful, interesting
and unique about mobile learning. They also start to
tackle what is perhaps the biggest challenge posed by mobile devices in the
space of learning — turning consumers into producers.

Mobile devices have many affordances, they have come at a cost to date.
Limited methods of input, storage and many so– called "walled gardens"
(heavily restricted methods of software distribution) have made smartphones and tablets ideal media consumption devices, but often less suited
to production. Empowering the next "creative class" necessitates breaking
that barrier and insuring that young people have opportunities to consume
great media, but also have the opportunity to produce it.

Again, the authors in this volume start to tackle that problem, both in the
ways that extend the experience beyond the software on the screen and
allowing students to create that software themselves.

The best days of mobile learning are still ahead. I fully expect this modality
to move closer to the center in the coming years, and the authors in this
volume will lead the way.

◉

# preface/foreword

*by Kurt Squire*

Ten years ago, Roy Pea and Jeremy Roschelle predicted that the story of mobile media (called handheld computers at the time) would ultimately come down to this: Will educators confront scenarios of radical decentral- izing– where learners might pursue passions and interests in an entirely uncoordinated way (and somehow coordinate them) or will the availability of these massive data exhausts that emanate from our lives– captured through portable devices– create Orwellian scenarios in which administra- tors know everything about students' lives?

This story is still unfolding, but clearly these are themes that educators wrestle with. What excites me about the work in this volume is the innovative ways that educators are wrestling with these challenges. Place and story can hold together groups of learners and be the springboard for coordinated activity. Learners can take control of their own data, and become more empowered participants in their own learning. Although we still are working on these issues, as a field, we've come a long way since those early forays with handheld computers.

The story of how this book comes about reflects these same tensions. Over the past decade, groups at MIT, Wisconsin, and Harvard have been discovering new ideas, trying out radical innovations, and coming back to share our work. This decentralized, mucking around has been held together by a sense of common purpose, good will, and a lot of trust, and watching this group develop, evolve, and now begin making an impact on the world is a real pleasure to see.

I might trace the history of this book to a conversation among Walter Holland, Eric Klopfer, Philip Tan and myself at MIT, where we were imagining the future of educational games (others would have their own ways of tracing the story). Could we build educational games that used the real world as a game board, using digital devices to layer a fictitious world around it? For me, this idea came to fruition when Jim Mathews designed Dow Day as a class project, suggesting to me the pedagogical potential of this technology.

I vividly recall watching Gunnar Harboe, an MIT Undergraduate trying to get a GPS fix on his hacked together PDA, and thinking, "No way that this is in schools within the decade." Within that decade, we now can point to students *making* games for such devices, which are becoming nearly ubiquitous. There's a lesson in Moore's law here worth reflecting on.

This book contains a lot of stories of people doing amazing things. Most of them were accomplished by people doing unusual things. Microsoft investing in learning technologies on a long−term horizon. Eric Klopfer collaborating with Henry Jenkins at MIT, and then with Wisconsin and Harvard, and each having the trust and goodwill to maintain a solid collaboration over multiple years and projects.

Two small, unusual, but critical things that occurred at Wisconsin were: 1) Mark Wagler joining our team, bringing decades of experience as a master teacher experienced in place–based and critical learning, and 2) Collaborating with David Gagnon and DoIT to run ARIS through central University IT rather than a research lab. The thinking behind this was, "If we run ARIS through University IT, maybe the University will institutionalize ARIS, and maybe other Universities (with a similar IT infrastructure) will be able to adopt it as well.

As we watch the number of mobile games grow exponentially, I'm confident that this was the right decision. "Letting something go" meant losing any semblance of control over it, and the trade–off has clearly been worth it. I hope that readers sense the spirit of inclusion and collaboration that pervades all of this work (which I trace in no small part to Eric Klopfer), and feel welcomed to join by playing, or making mobile media learning (or art) experiences on their own.

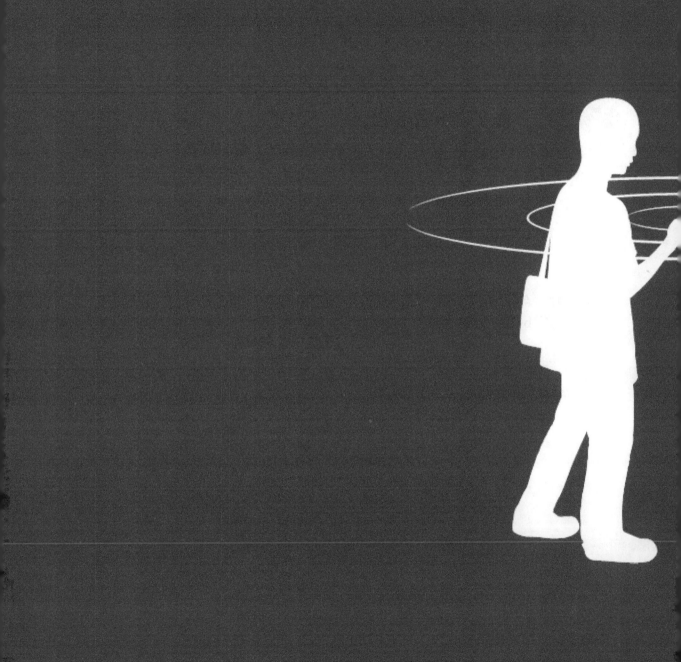

# table of contents

## build your own

## join the mobile media learning community

boot up

# dewey buys a smart phone

*by Seann Dikkers*

# dewey buys a smart phone

*by Seann Dikkers*
*University of Wisconsin — Madison*

## MOBILE MEDIA LEARNING

New mobile media technology is moving both information and communication capacity away from central repositories and into each individual learner's hand. Mobile media learning includes the instant and ongoing connection of hand–held devices to online information and communication for personal growth and increased agency within professions and communities of practice.

Learning about or how to do anything is no longer necessarily relegated to scheduled times or places, but can be accessed personally anytime, anywhere. Yet because the devices are both relatively young, the ways in which to best use these tools for learning is still in a compelling state of experimentation and design.

Also, because mobile devices are becoming more and more ubiquitous, while the tools of production are simple, powerful, and increasingly free; everyone seems to be doing something with them. This isn't a slow burn of adoption, it's more of a bonanza. Those interested in both informal and formal learning are left to wonder: What mobile media learning efforts show promise? Which can they do themselves? And most importantly, where are potent examples and models of mobile media learning?

This book shares a collection of stories where mobile technologies have been successfully used for learning. We make the claim that these cases represent more than just creative lessons; they represent potential for new ways to design learning environments that can be replicated, refined, and polished for use by you, the reader, this weekend.

Each chapter in this book features a purposefully selected educator from a variety of learning contexts and content area design goals — yet each leverages the power of intimate, instant, and invigorating learning on the move. Mobile media learning. Ideally, you should be able to look through the table of contents and find a chapter where mobile media has been used with skillful design in your context. We hope that you can be inspired to modify and adopt one of these examples and start on a road to mobile media learning.

Without intentional design and purposeful presentation of effective learning, coming changes may not make the most of the new technology available to educators. We argue that educators are in the best position to see what works for learning, not political figures or corporate think–tanks, but those creating spaces for learning. They can see immediate results, get needed feedback, and make changes quickly to leverage design for learning the most effectively — they can learn as they construct.

Before specific cases of mobile media use, this chapter attempts to briefly place experiential mobile learning into a techno– historical perspective. If the means of information and communication shift, so will workplace tasks and skills, and learning will soon follow. This context helps us to see the work of iterative design as a central task for educators today, that will have lasting impact if only their knowledge is proliferated, popularized and integrated at scale. If we want to leap into the future, we are well propelled by looking into our past — if only for a few pages.

## WE HAVE BEEN HERE BEFORE.

During the industrial revolution, the Prussian model of a 'factory school' inspired many. The idea that schools could be invented that served both the rich and poor at scale was a noble goal. Industrial giants saw education as job training. For a market economy of workers and consumers, workers needed to be familiar with an eight hour shift, standardized breaks announced by a bell, set eating times, and the capacity to stay in a single location for the durations between. Consumers needed to be ready to consume lessons taught and buy them in the market. But were these past goals humane? Good?

Today's schools still reflect the design solutions that successfully met the demands of the workplace and market then. But the workplace is changing, and with it a call for educational change. But are the new demands of the workplace the best designs for learning? Humane? Good?

Participation in the culture is finding validation outside of the schoolhouse walls and designs for learning need to address more than just workplace skills, but civic engagement, informal learning, production, and purpose in the lives of learners. Learning is something more essential to the human condition and is well served by an educator's perspective.

Since the changes of the industrial revolution, it's interesting that the voice of John Dewey has sustained conversation about individualized, constructivist, humane learning. He often asked questions that went beyond, "How do we prepare students for workplaces?" But are his questions still relevant today? For instance, what would Dewey have thought of mobile devices? How would he have thought to use them for learning?

Let's say he were to see a mobile device and saw the ways learners could play, create, build, and design with new, personal, mobile, technologies. Would he see the potential of individual learning? Interest– driven learning? Place– base learning? Data– collection? Inquiry? While during Dewey's life technology was centralizing and polarizing, mobile media individualizes and connects in ways that would have, and do, foster constructed learning. Dewey's progressive schools would have been well served with such technology. His constructivist vision would have had affordable tools to deliver learning at scale. I argue that Dewey would soak up today's new media and new mobile technology and quietly whisper, "Yes, finally."

## EXPERIENCE AND EDUCATION

However, Dewey had a healthy skepticism of technology. He did understand that student learning would be enhanced if it was situated in, and grew from, locative activities that learners themselves thought were relevant and interesting, but that didn't necessarily have much to do with technologies. In Experience & Education (1938), Dewey noted a need for experience based learning rooted in life beyond the classroom, yet "some experiences are mis– educative... that has the effect of arresting or distorting growth" (p 25).

Through Dewey's lens of Experience and Education, Dewey outlined an educational model that resembles closely what is being called for by the 21st century learning community. He called for an experience– based model for learning that included a study of exceptional practices (p 51); the premise of this book. His model called for teacher as judge of worth– while experiences (p 49), leader of group activities (p 59), and to take advantage of moments of revelation (p 71). Students should be guided in social skills, self– control, and observation (pp 62– 64); towards intelligent actions (p 65), connected to the community, growing environments, organization of facts, and purposeful activities (pp 73– 88). Today's mobile media learning resembles Dewey's vision more than not.

Though there are still potentially mis– educative experiences, according to Dewey's definition, there is also growing potential for learning outside the walls of the classroom too. Mobile technologies today facilitate locative learning without the "whim and caprice" (p 65) of setting young learners loose in the world without direction. In this light, Dewey would have bought a cell phone for himself, designed new experiences, and potentially been a writer for this book — or at least it's fun to think it so.

## WHAT WOULD DEWEY DO?

Would Dewey be stunned by the amazing progress that has been made? Would he also see largely the same personalities in the government and business sectors today as he did then? Are there echos of those pushing for more and more centralization of control, accountability, and standardization of the learning process? I like to think Dewey would have been pleased to see many still working for student– centered instruction, inquiry, and experiential learning by designing alternative models, charters, and ongoing innovative work.

As in Dewey's time, brand new educational professions are emerging (like 'aides', 'special education', 'instructional compliance staff', 'mentors', expanding administration, and even police officers) in the school environment. Dewey would see attendance and drop– out rates, college admission rates, the rise of Montessori and Homeschooling; and formal schooling in a period of change. Learning itself is changing and the surrounding systems are straining to adjust.

There is no doubt that Dewey would feel familiar with the buzz over new technology — primed to change lifestyles, and already causing waves of profitable reforms in business, entertainment, and public policy. Dewey claimed that experience is both the means and end of education. The experiences we design are for the core of learning. For instance, testing may be good in some ways, but it can potentially dictate the entire experience of education if allowed. Is the 'ability to test' the core experience we want for our youth? Education is the experience we design. Likewise, what do mobile media provide in terms of experiences they generate?

Some seek to put computers in every school (done), interactive whiteboards in every classroom (getting there), laptops in every lap (spoken of), and soon we'll hear call for hand– helds in every hand. But do these tools equate to learning experiences? Dewey may see a corporate influence on education that simply buys its way into the 21st century or a bandwagon effect that has a similar force and rush as the industrialization movement of his time — and has clear financial benefits for those selling the tools ($1 Billion in Interactive White- board Sales by 2008). Yet do all of these efforts amount to stronger education for our youth? Do these technologies have a cumulative or direct effect on learning? Is new media making us smarter? These efforts are rooted in an assumption that provision will have an impact on learning.

The community of educators that have chosen to participate in this book have each found ways to design first in order to understand through experience and practice. We seek to understand mobile technology for learning prior to any call for mass provision. What applications, experiences, spaces, communi- ties, problems, verbs, collaborations, and designs will best leverage a marriage between already rich real– world spaces and technologies that serve, Dewey– like, to focus, direct, and provide a lens for these experiences? Well, here are some amazing cases of practice — each showing new elements and visions for a new kind of locative learning mediated with technology.

In the following chapter we use Dewey's targeted questions as a template for the coming cases of mobile media learning.

# asking experts
# dewey questions

*by Seann Dikkers*

# asking experts
# dewey questions

## by Seann Dikkers
*University of Wisconsin — Madison*

We decided to make the design of this book transparent. Below you'll find detailed descriptions of the questions we posed to our authors. Not only did we want to get the most out of the expertise of the authors, but we wanted to let you in on the process too — for two reasons.

By showing the prompts for the cases, you get a sense of what questions the authors were wrestling with and can see how they framed them. These questions are the only real common ground this book has as each author's chapter has their voice, language use, and even distinct advice to you the reader. At times our experts don't agree, and we love this. Notably, you'll see as much creativity and ongoing differentiation of practice as we did — but the questions were a common starting point. We see these differences as positive. In fact, we wouldn't want a set of common questions that led to stock answers. We have so much to learn and great questions lead to great answers even if they diverge some.

Also, to put it bluntly, you may be a future author for this book. ETC Press has a fairly innovative publishing model that allows the editorial team to add or update chapters for each publishing of the book. Mobile media learning is still so new, our authors are currently on their next projects, and if this book inspires new practitioners, then we believe this book can and should grow with its community. If you give mobile media learning a go and find it inspires powerful learning experiences, then the questions below save us a lot of time before you start writing.

Using Dewey's questions, we've structured this book to provide a collection of early suggestions based on field– tested learning experiences using mobile devices. In Experience in Education, Dewey asked six key questions for designing education:

- What does traditional vs. progressive education look like now?
- What experiences are valuable for the learner?
- How should we address social control in education?
- What is the nature of freedom for learning?
- What is the meaning of purpose for learning?
- How should we reconsider the organization of the subject matter?

You too can consider the questions for yourself as you read each case and possibly return to them when you write your own chapter.

## QUESTION 1
## what does traditional and progressive education look like?

What is being done in and around teaching and learning using mobile devices? How might this look? That answer is different for each design effort. In the years to come, models of 'what works' can merge with a philosophy of 'valuable' learning and these will be the path forward for education. We celebrate that many outstanding potentials are already in play and seek to show, not tell, what education with mobile devices may look like.

Given the newness of the technology, we have gathered stories that show potential in addition to enthusiasm in multiple settings. In each case the researchers and educators concluded:

*a) There were powerful possibilities waiting, and*
*b) Mobile facilitates entirely new ways to think about teaching and learning.*

The following chapters present mobile devices, used as learning tools, in both formal and informal environments, indoor and outdoor, with and without teacher guidance, and at times have students playing, searching, working, and designing for themselves. Notice that each case has already seen learners use the technology so we can see, physically even, what learning looks like.

## QUESTION 2
## what experiences are valuable for the learner?

Dewey's second question provides clear direction for each of the writers to consider their designs from the learner's perspective. Value for the state and value for the learner are essentially different. Common standards have never been good at considering the individual, but mobile technologies thrive on individual faculty.

This question artfully redirects our attention from test scores, to learning; considering what experiences are valuable for learners. If the experience isn't, at least, considered valuable by the learner or by the local teacher, than no amount of effort, energy, polish or design will convince teachers and school leaders to adopt new practices and policies or change the current system.

In each chapter we ask for the writer to give some sense of the learner's feedback on the experience itself. The perception of value doesn't equate to measurable outcomes. If I see an activity as relevant, I may be more engaged, active, and learning at an increased rate, but we aren't any closer to knowing what I'm learning, how I'm learning, nor how much I'm learning. Though entirely relevant to constructing my learning, the degree to which I value an experience isn't neatly measured and tallied. The only way to measure this is to simply ask the learner, "Was this valuable? How?" and know that this is an essential part of the problem of teaching and learning.

## QUESTION 3
## *how should we address social control in education?*

Dewey makes the case that education is an agent of social control. The goal of an education system is, in part, to direct and guide a national level of behaviors and competencies that both serve the society and protect it from harmful behaviors. Education helps train a citizenry for citizenship as defined by those designing by and paying for the system. We ask how mobile learning prepares learners for citizenry in a larger community.

We address local management of learners to some degree too. No educator likes the idea of a class of students running about all willy–nilly — especially outside the walls of the classroom. Dewey acknowledges and sets aside both those that worry too much, and those that are too permissive. There is a balance between control and license. Learning in any formal setting needs to be organized with better planning, ongoing improvements, careful organization of the contexts used, and reviews of growth and play within the activity — even with mobile devices.

Dewey clarifies that the teacher's role of guide and leader is not that of being the 'boss' in a workplace sense, but of being a mentor and organizer of activity. The teacher should plan ahead, respond during, and review after every experience they orchestrate. They also should value the activity and their role modeling social skills, problem solving, and enthusiasm to students. Even in informal settings, the teacher isn't reduced to administration, but is central to the design and creative application of an experience locally; teachers are central to these experiences.

We ask writers to reflect on the process of planning for, guiding, and reviewing their activities with mobile devices; and consider the role of the teacher. In reading each section, educators will present an instructional/organizational guide for practice. Knowing that all activities will need some localization, you should be able to see how and what is to be done.

## QUESTION 4
### *what is the nature of freedom for learning?*

Mobile is mobile. Student movement and use of mobile devices can be leveraged in ways that Dewey only dreamt of. Even today we are still trying to nail down what are all the ways possible that the learner can interact with the spaces. The question here isn't whether or not mobile devices increase freedom, but to what degree and what shape should that freedom take?

For Dewey, it was important that a learner at least perceived that their learning was their own — even if they understood it was part of an overall compulsory system. This meant that learning was somewhere between the extremes of 'whim and caprice' and clearly false control. Students should neither be allowed to waste away their time, nor be given false options (i.e. Do this work, or you're free to go to the office).

The cases in this book capture a range of applications of mobile devices in terms of freedom for the learner. Some are compulsory classroom activities and others are purely voluntary experiences. Every decision of choice is an effort to strike a meaningful balance between control and permission. Explore the chapters considering your own comfort with learner freedom and how to design with a balanced eye for learning.

## QUESTION 5
### *what is the meaning of purpose for learning?*

In terms of mobile, we find again and again that new experiences call into question the very meaning of purpose for learning. Both the learner and the educator should embrace the importance of purpose in the experience. Youth bounding through the woods, doing quests in the city, or even re–creating their profile to maximize their babysitting time — these are clearly different from traditional learning: How does 'purpose' play into the activity? Where do the experiences lead? What new opportunities do they open up?

When new tools give leverage in the larger community, schools are often asked to provide students with experiences that allow them to gain proficiency with those tools. For instance, if the capacity to read is relevant, than schools teach reading. If the capacity to compute basic math is relevant, than schools teach math. If the ability to quickly connect, network, and produce ideas is relevant, than there will be pressure to teach using mobile devices and computers.

When students ask, "What is the purpose of this," and we should be willing and able to answer. In fact, various organizations are claiming that there are 'new media skills' that every student should have exposure to. Designing with purpose allows for new designs to address those needs, while at the same time considers the importance of lifetime learning. The purpose of mobile may not neatly fit into reading and math scores, but should they? Or are there other valuable purposes that are being addressed?

Dewey challenges us to never allow industry to define the entire context of learning and growing as a human being. Industry may give a nod to music, art, creation, and innovation, but will never place it in the forefront; because, in essence, they are then promoting potential competition for consumption. Educators uniquely can ask what purposes best serve the learner.

## QUESTION 6
### how should we reconsider the organization of the subject matter?

Dewey didn't assume that segregated subjects constituted necessary distinctions for 'content'. Content was more broadly defined as 'life experiences' that the organization of curriculum provided. The researchers and educators presented here have already begun to ask and find answers to how mobile can activate and enable 'content' area learning. They have designed for, around, or despite traditional learning organization while seeking to be educators. The work presented here ,even the informal examples, all represents efforts to convey content of some sort. These are learning models.

Each of the following chapters approaches content in one of three ways.

1. Mobile devices give students access to traditional subject areas.
2. Mobile devices open up potential for meaningful interdisciplinary activity.
3. Mobile devices open up entirely new topics of learning and skills needed to learn.

Dewey also considered locative experiences to be valid educational experiences. Writers often consider the larger picture of education and place their work within a larger context, namely because mobile learning models offer potential to change learning models. Because this isn't an attempt to construct an academic piece we give these outstanding thinkers a chance to consider "What if, what may, and how could education look in the coming years?" and ask them to share their ideas freely. We do not offer a proposal for reorganizing curricular content, but we do suggest this is ripe for discussion.

If Dewey had a smart phone, I believe he'd love this project too. This book is a collection of examples of progressive mobile research and projects conducted in the field, across the country, and with learners. We have gathered leading educators and researchers, that are already asking the above questions, to share with us and collectively build a window into early mobile iterations for learning. None of the authors would claim they have the answer to any of these questions, but we believe they are making headway on all of them through iterative designs for learning with mobile media technologies.

What will the future of learning look like? Here are a few possibilities.

◉

play to learn

up river: place, ethnography, and design in the st.louis river estuary

*by Mark Wagler and Jim Mathews*

# up river: place, ethnography, and design in the st.louis river estuary

*by Mark Wagler and Jim Mathews*
*University of Wisconsin — Madison, Local Games Lab*

As teachers, we typically investigate nearby places with our students, and together design interactive stories intended for local audiences. As part of a larger research and outreach project, we built on these classroom experiences to design Up River, a mobile story and an associated two– day workshop — intended especially for teachers and students interested in working as ethnographers and designers to explore and represent their own local places.

In Up River, players travel upstream from the Duluth harbor in search of wild rice and native fish species. Along the way they become physically immersed in the estuary, exploring tourist attractions, industrial sites, restored habitats, and fishing piers. To complement these real– world experiences, they also access geo– located stories and science via their mobile devices. In the workshop, participants engage in mobile design activities that can be further developed back in their classrooms.

*duration:* 2– 3 hours
*location:* In the estuary formed by the St. Louis River as it flows into Lake Superior at Duluth, Minnesota and Superior, Wisconsin.

## PLACE

Place is the core of Up River — it is the context for collaborations and designs, subject for inquiries and documentation, locus for hands– on and virtual experiences, and a complex system whose dynamics are progressively revealed through the storyline.

We designed Up River by immersing ourselves in the estuary. As we headed upstream from Duluth, then back downstream on the other side towards Superior, we probed every possible access to the St. Louis River. We explored boat launches and beaches, parks and forests, backwaters and rapids, camp-grounds and historic sites, and many wharfs and other private businesses — if there was a path to the river, we followed it. During these field excursions we met a variety of people connected to the estuary; at the same time we began identifying locations and stories that would later help us immerse players in the same environments.

### st. louis river estuary

The St. Louis River flows 179 miles and drains a 3,634 square mile watershed. The lower river opens into a largely wild, 12,000– acre estuary before it finds its outlet in the highly industrialized Duluth— Superior harbor, the world's largest freshwater seaport. Here's how we describe the estuary at the begin-ning of Up River:

The St. Louis River is the largest of the U.S. rivers that flow into Lake Supe-rior. As it nears the lake, the river broadens out into an "estuary," or "drowned river mouth." The estuary is a mixture of water from the river coming down-stream and water from the lake moving upstream.

A variety of waterfowl and other wildlife use the estuary for breeding and migration including about 54 species of fish (e.g. lake sturgeon, walleye, yellow perch, northern pike, and black crappie).

### locations

Up River takes place in three major locations, each presenting a unique view of the estuary. While water, resources, and contaminants flow downstream, the narrative guides players upstream — from a tourist district, to the hub of shipping and industry, and finally to a restored wetland. The story begins in Canal Park, once a warehouse district, and now the center of Duluth's entertainment industry, with restaurants, hotels, shops, and tourist attractions. Here a local chef sets the stage by giving players their main quest:

> *Hi there! I'm hoping to make a meal tonight based on local ingredients and I need your help. While a lot of restaurants here in Canal Park serve walleye, most of it comes from Canada. Also, the "wild" rice they have on the menu is usually commercially raised. I want my meal to be as local as possible, so I need you to bring me fish and wild rice from the St. Louis River estuary. You can start by collecting information here along the Duluth harbor, but you'll need to head upriver in order to complete your mission.*

As the players explore key sites along the St. Louis River, they meet a variety of contemporary people, both real and fictional, who live and work in the estuary. In Canal Park, they interact with a fishing guide, ship watcher, rice vendor, and several recreational and commercial fishermen. They also meet historical figures, such as Henry Schoolcraft, a 19th century geographer and ethnologist, who shares original journal entries describing the harbor area in the early 1800s, long before it was altered by large scale European settlement and industrialization.

Some people appear on the players' mobile devices as virtual characters, while others are real people who agreed to serve as prearranged interview subjects. Some of them serve as guides and help players complete sub– quests, while others simply share their musings about the estuary, or even attempt to distract players from their main quest by encouraging them to engage in side activities, such as watching a cargo ship entering the harbor, eating at a local restaurant, or doing some sightseeing.

Unable to catch a legal– size walleye, or purchase "real" wild rice, players must travel several miles upriver to the second location, Rice's Point, where they encounter a dramatic change in environment. Here in the center of shipping and industry in the Twin Ports, surrounded by ships and docks being loaded and unloaded with cargo, players can look across the deep river channel to Connor's Point in Superior, or just a little upstream to the sewage treatment plant. A new group of virtual characters await them — a fisheries specialist, a boater, an invasive species researcher, and a conservation coordinator with the U.S. Fish and Wildlife Service — as well as documents exploring the history of sawmills at Rice's Point, the impact of the sewer plant on water quality, and the invasion of zebra mussels.

Still searching for that elusive wild rice, as well as additional fish species, players must travel even further up river to their final location, Grassy Point — a restored wetland representing what the river might have looked like before industrial development and heavy dredging. Located next to the farthest upstream wharf still in operation on the river, this was once the site of several lumber mills. While the mills played a central role in the industrial and economic development of Duluth, they also caused severe ecological damage. As players explore Grassy Point, they learn about the industrial past through photos and historical accounts, but also about the efforts to restore the area and improve it as a habitat for fish and birds. It is here they encounter John Turk, who shares a story about gathering wild rice just across the river as a child in the 1930s, several birders who help players identify different bird species, and an angler who describes how the restoration improved fishing in the area.

### systems

While exploring Up River, players learn about the natural ecosystems of the estuary. They learn that wild rice won't grow in deep or swift water; the round goby is an invasive species introduced through the ballast water of ships; restored industrial sites, like Grassy Point, support a range of aquatic plant and bird species; and decomposing waste uses up oxygen needed by fish. More significantly, they experience the estuary by walking on the shore of Lake Superior, standing at the end of a pier above a deep river channel, and exploring the wetland where Keene Creek enters the St. Louis River.

Players also learn about the cultural systems of the estuary. They walk past tourist attractions such as the Lake Superior Marine Museum, see charter fishing boats and 1,000 foot cargo ships travel through the shipping channel, explore a boat launch in the middle of the industrial section of the harbor, and walk past an active wharf with enormous piles of dry bulk commodities. On their mobile devices, they interview virtual fishermen and dock workers; shop from a rice vendor; and go fishing. These interactions, and the stories players hear, help connect them with the people and places in the estuary.

Most importantly, Up River draws attention to past and present interactions between the cultural and natural ecosystems in the estuary. At Rice's Point, for example, players are asked to consider the connection between natural resources flowing down the river and sediments and contaminants flowing into the estuary. The following text accompanies an historic photo of a large log flotilla that was taken at Rice's Point:

> *The lumber industry dominated Wisconsin's Connors Point [the point directly across the river from Rice's Point] between 1860 and 1909. White pine logs felled in the St. Louis watershed were floated downstream. By 1894 at least fifteen sawmills were located along both sides of the St. Louis River. These sawmills, along with iron works, shipbuilding yards, and other industries, introduced a heavy load of contaminants into the estuary.*

The logging boom ended quickly so that by 1925 only one sawmill remained in Duluth, and the white pine forests, which once seemed inexhaustible, had disappeared. After the trees were clear cut (deforestation), the rainwater that once infiltrated the forests now runs off more quickly, carrying eroded soil with it. Much of the resulting sedimentation ends up in the lower St. Louis River, degrading the ecology of the estuary.

Learning about the estuary through the stories of people who live and work there helps players "see beyond" what is in front of them, making transparent some of the complex systems at play in the estuary.

## ETHNOGRAPHY

Ethnography is central to Up River. Not only did we conduct research using ethnographic methods, we designed the story itself to guide players in practicing simple ethnographic techniques. By ethnography we mean everything from people watching and conversations in everyday life, to the systematic observation, recording, and analysis of cultural phenomena undertaken by professionals in many disciplines.

Up River represents a distinctive genre of mobile storytelling. Instead of relying solely on fictitious or historical characters, it emphasizes contemporary real people whenever possible. Up River also encourages players to observe, interview, and record real people, places, and interactions. Indeed, Up River served as a model during our workshop for how readily ethnographic documents can be incorporated into a mobile story. Many teachers were intrigued by this approach and easily began generating ideas for how they could conduct ethnographic research with their own students.

### ethnographic fieldwork

Our first step towards designing an ethnographic story rooted in place was to immerse ourselves in the estuary. As we visited multiple sites up and down both sides of the river, we took hundreds of photos and conducted extended field observations. This gave us an entree into the rich culture of the estuary and provided a frame of reference for the many conversations we initiated during repeated visits to our core locations. By referencing what we were learning about local places, practices, people, and events, we were able to further immerse ourselves in the estuary and develop even more nuanced questions and understandings. We slowly discovered a local network of people who live, work, and play in the estuary; most were eager to share their experiences and ideas.

Since many of the people we talked with were fishing, we began thinking of how to include fishing in our mobile design. For that we needed more information. Web resources for this geographical area are bountiful and answered most of our basic questions (e.g., related to fish populations, water quality, invasive species, commercial fishing). We were also grateful for interviews conducted by landscape ecologists on our research team and the vignette they wrote about fishing for our project web site. What we were missing, however, were personal stories tied to the specific places we wanted to include in Up River. For that, we needed to conduct longer, recorded interviews in the field.

One of our favorite locations to conduct field interviews was Boy Scout Landing, a boat launch just above the historic Oliver Bridge, some twelve miles upriver from Lake Superior. The small landing, which includes a fishing pier, boat ramp, and tiny sandy swimming beach, is on the Duluth side of the river, between the mouth of Sargent Creek and River Place Campground. Across the river, on the Wisconsin side, the land is undeveloped — with wetlands in the foreground and forest behind. In addition to providing easy access to the river, Boy Scout Landing is also a great place to meet anglers.

On one visit we met a young man who lives in the neighborhood, fishes at the landing, and used to swim there. He told us that it is mostly local people who fish on the pier, except during the spring walleye run, when the fish head upriver from Lake Superior to spawn — "on opening weekend, they come 20, 30 people on the dock."

As we talked more, he told us that he catches a lot of bass and catfish in the river, as well as some invasive species like white perch, gobies, and roughies, but then added, "I really like fishing for northern and muskie, they're more fun to catch." This sparked him to show us some of the lures he uses, including a top– water spinner that works well in shallows because the vibrations attract fish, and a floating rapala that moves side– to– side and works because "it looks like a dead or wounded fish." We also talked a bit about sturgeon and he pointed to another fisherman on the dock who caught a fifteen– pounder. But since they are endangered, it's illegal to keep them, so they unhook them as fast as they can.

When we asked him what he does with the fish he catches, he told us that he eats some, but mainly catches them for sport, then releases them. "A lot of mercury in the water, and they say the older the fish and the bigger they are, the more mercury they contain." Before we left, he told us that fishing has gotten better in the area and that he sees fewer diseased or unhealthy looking fish — something that was common in the past.

Extended field interviews like this helped us develop a feel for the people and places in the estuary. We conducted many of them, in multiple locations along both sides of the river. In the end, they were invaluable because they provided ideas and content for our design and made it possible for us to more accurately present local people, stories, and issues in the final version of Up River.

### ethnographic content

One of our main design goals was to give Up River an authentic real–world feel. Our extended fieldwork and documentation helped in this regard because it allowed us to build our narrative around site–specific media and stories. We also used photos and videos we took in the field to more accurately "place," or situate the final story. For example, players see Mark Howard, a commercial fisherman, at Howard's Fish House & Farmer's Market where we interviewed him. Wearing his white apron, and talking with expressive gestures, he describes one environmental threat to his profession:

> Most of the effects on the fishing that have happened here have been through the introduction of exotic species, the smelt, the salmon, zebra mussels, the gobies, the roughy fish; and they have competed with the native fish for food sources and habitat and eaten the fish.

> Smelt are carnivorous; they are the most devastating thing to ever come here... They ate all the baby white fish and herring and the walleyes and the perches that were there in the bay. They would come in en masse and just wipe out everything when they spawned.

Similarly, we used a picture of Don Nelson's charter fishing sign to help players find his boat in the harbor, an image of a restaurant's menu to clue players into the source of the food they serve, and a link to a Facebook page depicting a local organization's effort to clean up Grassy Point. We also took photos of people standing in the same location where players encounter them in the story as a way to connect virtual interviews to real–world places.

## present– focused ethnography

While stories from the past can be found in Up River, the narrative emphasizes the present–day practices, beliefs, expressions, and social fabric of the local culture. Even when stories about the past are used, they are framed in a way that draws us toward the present. Consider the following text, transcribed from an audio clip that appears in Up River, excerpted from a much longer interview conducted by one of the landscape ecologists on our project:

> There was some huge beds of rice on the St. Louis in the early 30s. And I did a lot of wild ricing in the 1940s, late 40s, and into the 1950s, and that's when it started to disappear. But all these bays above the Oliver Bridge, and below the Oliver Bridge, were full, full of wild rice. Big Pokegama Bay, Allouez Bay had a lot of rice, every bay here was loaded on both sides, the Minnesota side and the Wisconsin, there was more rice on the Wisconsin side than the Minnesota side. Every bay here was loaded, was loaded with wild rice.

When players encounter the speaker, John Turk, a lifetime resident of Oliver, Wisconsin, they get more than bare historical facts. Through a contemporary storytelling style — the intimacy of his voice, his cadence, word choice (e.g. lots of dates and place names), and especially his rhetorical patterns (e.g. the accumulative impact of repetition) — John communicates his present and deep connection to the estuary. This is not an abstract past, but living memory passed to the present. His voice sounds authentic, his personal experience lends authority. Additionally, while John's story depicts wild rice harvesting in the 1940s and 1950s, the narrative in Up River encourages players to ask questions about current cultural practices surrounding wild ricing.

### ethnographic nudges

We want players to observe and to interview. Thus we designed Up River to kindle attention to and interaction with the local environment. Early on, players look for native fish at the Nels J., a former fishing boat turned seafood snack stand on the shore of Lake Superior. Using a dialogue script, players ask: "BTW … what's the story of the Nels J.?" The attendant replies: "The boat was once used to fish on Lake Superior. You can learn more by reading the posters on the kiosk right in front of us." Instead of placing the text and photos from the kiosk in Up River, which would focus their eyes on their mobile device, we use character prompts to guide players' attention to their physical surroundings.

Similarly, after learning about the industrial past and subsequent restoration of Grassy Point, players are given a set of questions nudging them to observe the current activity at an adjacent wharf:

> While the sawmills are no longer here, there is still a lot of commercial activity in this area. How has commercial land use changed over the years in this location and why? What do you think comes and goes out of the wharf just to the west? What are those big piles? Where do those resources get used and how? Check your inventory for a wharf report that might help you answer some of those questions.

This prompt, along with the additional details delivered via the wharf report, encourages players to notice and ask questions about the various materials being loaded and offloaded there. It also reinforces what they learned moments earlier about the actions taken to restore Grassy Point from a contaminated industrial site to a relatively healthy wetland.

We also nudge players to talk with and even interview real people. Immediately upon arrival at Rice's Point, for example, players receive a phone call from the Chef with an ethnographic assignment — interview live people at this site, and record these interviews (and perhaps take photos) with your mobile device:

*Just heard there are several people out at Rice's Point right now you'll want to talk to. They might be walking around, so use their photos to find them.... Look for Mark Howard, a fisherman, and Pat Collins, a conservation coordinator.*

*So we can remember what you find out, I need you to document it.... Tap the "MORE" button at the bottom of your screen to use your digital recorder to capture audio interviews. Short clips! Save your battery! There's a camera, too, if you want to show me some photos.*

Up to this point players have only conducted interviews with virtual characters by selecting questions from a list and reading pre– scripted answers. In case some players might hesitate to approach a real person, Up River makes the task easier by suggesting kinds of questions they might ask:

*Find Pat Collins. He's ... with the U.S. Fish and Wildlife Service. Ask him if it is OK to eat fish you catch at Rice's Point, or is the water cleaner and the fish healthier if you go farther upstream. For sure he'll tell you how WLSSD, the local sewage treatment plant, has changed water quality in the St. Louis River Estuary. Also ask him about habitat restoration in the estuary and whether it has improved fish spawning and the availability of wild rice.*

## DESIGN

Design pulls together everything we worked on across the project. We planned, built, and refined Up River and the associated workshop by continuously highlighting place and ethnography. In this section we present a few of the additional goals, opportunities, and strategies that guided our thinking as we designed Up River.

### participatory design with partners

Up River was designed as part of a larger research and educational outreach project, "Stressor Gradients and Spatial Narratives of the St. Louis River Estuary," in which a multi– disciplinary team gathered, analyzed, visualized, and disseminated scientific and cultural data.

Limnologists and aquatic ecologists at the University of Minnesota — Duluth and University of Wisconsin — Superior sampled multiple estuary locations to determine the relationship between human impact (via sediments, nutrients, and contaminants) and environmental indicators (water quality, wetland plant and macroinvertebrate communities). Landscape ecologists from the University of Wisconsin — Madison and Bemidji State University combined this data with ethnographic interviews to create a series of vignettes — developed around five core themes (shipping, fishing, mining, recreation, and wild rice) — to represent human— ecological relationships in the estuary. Outreach specialists and educators from Wisconsin and Minnesota Sea Grant developed educational materials and programming.

Our role, at the Local Games Lab of the University of Wisconsin— Madison, was to design a mobile story playable in the estuary. Throughout the design process, we repeatedly tapped the knowledge, professional expertise, and creativity of our project partners, including people we met in the field and teachers and students at the workshop. Early in the project, we implemented a two– part design jam where we introduced our partners to mobile storytelling by having them playtest a prototype we designed for Grassy Point. After they provided feedback on our design (which was later incorporated into the final version of Up River), we asked them to brainstorm and mock– up their own story set in the estuary. In teams of two or three, they quickly came up with story ideas and components on sticky notes, arranged them into a narrative, and reported out to the group. The initial, rough narrative for Up River was born at this session.

### selecting locations

We located Up River across three key places, a minimum we thought for exploring the many aspects of a large estuary. We chose locations that would highlight the cultural and scientific research of our partners and pull together the five core themes of the entire project. We also wanted sites that would offer a variety of cultural and ecological features; provide rich sensory and cognitive experiences; and represent past and present uses and conditions. While seeking sites that were easily accessible, we also wanted to direct users upriver and away from the Duluth Lakewalk, a primary destination for visitors and school groups. In part, this was driven by our belief that in order to "experience the estuary" one needs to explore less– travelled areas that also exemplify its richness and complexity.

### flow

Our project team sometimes referenced "flow" as a unifying theme for the estuary. Natural resources like taconite (from iron mines), grain, salt, and coal flow in and out of the estuary in ships, trucks and railroad cars. Sediments and contaminants flow into the river through outfalls and tributaries, then downstream with the current. Meanwhile, lake water flows upstream as the result of a seiche effect, fish swim upstream (in the St. Louis and its smaller tributaries) to spawn, and the development of industries and settlement of neighborhoods historically flowed upstream from the harbor cities.

We integrated flow into our mobile story, and hence movement into our narrative, by having players head upriver to catch native fish and find wild rice. This was accentuated through the design, as virtual characters point players to sites or experiences further upstream — "you'll find better fishing up at Rice's Point" — and the quest structure requires them to complete tasks across multiple sites. In the end, players experience the river as it transitions from a heavily dredged outlet, to a fast moving channel in the center of the harbor, to a wide river with grass– filled bays.

### interconnected themes

As John Muir famously said, "When we try to pick out anything by itself, we find it hitched to everything else in the Universe." We designed Up River to help players better understand the estuary as a complex cultural and ecological system. To achieve this, we imagined storylines that highlighted one of our five "spatial narratives" — fishing, mining, recreation, shipping, and harvesting wild rice — and how each of these human activities stress or are stressed by the environment. Alone, however, this was not enough.

We discovered we could more readily create a complex representation of the estuary by combining at least two of these themes in our narrative. When we paired fishing and harvesting wild rice, the idea of finding ingredients for an estuary meal emerged — and we had the main quest for our players. It was easy, then, to bring in secondary connections to shipping and recreation. In retrospect, it almost feels as if Up River wrote itself, given how obviously the narrative fits the needs of our project — and triggers investigating the estuary. In truth, we spent a lot of time exploring locations and developing our own understanding of the estuary before finding a narrative that "clicked into place" some of the complexity in the estuary.

### humor

Sometimes Up River can feel as loaded as a Lake Superior freighter. Feeling a tension between providing rich details, but avoiding too much content, we regularly reminded ourselves that players can go to the project web site for more information. What they needed, particularly in the beginning, was a lighter touch to get them engaged. Enter the Chef — depicted as a cartoon cook, demanding and always in a hurry — who serves as the player's employer, the stage manager who hands out primary and secondary quests, and rewards players' success or failure.

After taking a moment to sample oysters from a fish stand, the player is immediately busted: "Hey, taking a break already? What are you eating? Oysters? Those aren't from the estuary. It's hard to find good help these days!" The nearby Ship Watcher commiserates, "Chef giving you a hard time? He fired me

because I'm not a work maniac like him. What's he got you doing?" When
the player buys "wild rice" from the Street Vendor, the Chef comes back
to scold, "Hmmm . . . looks like you got fooled! That's cultivated 'wild' rice
raised by farmers, not the rice that grows in the wild." Still, the Chef can
be kind and occasionally encourages players to keep their chin up.

## WORKSHOP

While teachers can use many approaches to cultivate student interest in
studying local places, a well– designed interactive mobile story or game
can engage, inform, and structure student field experiences, and lead to an
extended place– based project. We wanted Up River to immerse teachers
and students in the estuary, but also serve as a model for how they could
design a similar experience with their own students. Not only could Up
River serve as a hook to get teachers and students interested in studying
the estuary, it could also hook their interest in designing similar place–
based mobile stories in their own school or community.

We embedded Up River in a workshop, so that teachers and students would
have time and support for envisioning and planning their own place– based
mobile projects. Up River was first implemented in a two– day workshop
with eight secondary teachers from the region, fifteen of their students,
and some of the professionals involved in the larger project.

### debriefing sessions

Participants worked through the first segment of Up River while still
indoors in Duluth's Great Lakes Aquarium, accessing digital content
without experiencing its real– world context. When they next braved
November weather to play the final two segments outdoors, players had to
visit specific GPS locations in order to interact with virtual characters and
obtain other geo– specific multimedia via their mobile devices.

We facilitated debriefing sessions after playing Up River for two main purposes: one, we wanted feedback for future re– designs; and two, we wanted participants to transition from thinking like players to thinking like designers. Several students reflected that it felt "more real" and "more interesting" when they played out in the real world, because playing indoors is "not the same as being there." Participants reported several software and design bugs that interfered with their play, which we noted for possible fixes. Students in particular thought we should reduce the amount of reading, replacing some of the text with more video.

### designing with ARIS

We created Up River with ARIS, "a user– friendly, open– source platform for creating and playing mobile games, tours and interactive stories," available free at http://arisgames.org. We included a "how to" session in the workshop during which participants learned how to use the ARIS authoring tool in order to design their own stories. In this semi– structured learning environment, participants were guided through a basic design sequence, but then had time to ask individual questions and experiment on their own. As individuals and pairs completed their designs, they shared them with each other in order to work out technical issues and receive ideas for improvement. As is typical, several teachers and students quickly learned the basic functions of the software and emerged as experts, moving around the room to assist others.

### design jam

The pedagogical power of ARIS comes less from students playing stories designed by others than from students researching and designing their own stories. As participants began to develop a deeper sense of the core ideas — especially the confluence of place, ethnographic fieldwork, and mobile– based storytelling — we asked them to begin planning a similar project they could implement at their own school. Using a template we created, each school team brainstormed possibilities. As part of this process, we asked them to consider places, issues, and themes relevant to their community (and curriculum goals), as well as data, organizations, people, and any additional resources that might assist them in their design journey.

After reflecting on their ideas, each team picked one to develop further. At this point, they created a map and began to place people, items, and locations that might appear in a final mobile story. One of the teams developed a mock– up of a story exploring the history of a local running path. Another school, situated near the St. Louis River, presented an idea for studying their community and adding it as a new location to Up River. Other groups planned to investigate the effect of river pollution on human health, the impact of logging on natural habitats, the relationship between land use practices and water quality in a local river, and changes that have happened to a riverside town.

## LESSONS

As we discussed next steps with teachers at the workshop, we developed a set of talking points for how different school teams might be able to move forward given their specific needs, interests, and context. With this in mind, we end by sharing nine rules of thumb that emerged from the workshop, as well as from our own experiences as teachers and designers.

### *1. start small*

Create a simple tour before embarking on a more complex effort to design a mobile game or interactive story with lots of moving parts. Experimenting with many simple (and smaller) ideas helps build your fluency with the tools, clues you into what does and doesn't work, and often sparks ideas for more complex designs. Some teachers start by designing something for their personal use in order to learn the tools and experiment with different design ideas. Another strategy for starting small is building a tour or story around a single person or location, versus multiple people or locations. For example, imagine walking through a park with a biologist as she points out and discusses different invasive species, or touring a neighborhood with a resident who has lived there for 40 years as he tells stories about things that have changed; but now while you walk through the physical park or neighborhood, the biologist or resident you encounter are virtual characters on a mobile device.

### 2. start local

Build something that revolves around your school, schoolyard, or immediate neighborhood as a way to build your fluency with ethnographic fieldwork and mobile design. This allows you to experiment with the technologies and teaching strategies without having to organize lengthy field trips.

### 3. pilot your teaching strategies

Implement a small design project with a subset of students. In some contexts, this might mean running an after– school workshop or organizing a field trip for a group of highly interested students. Or you might organize a design competition, allow students to design a mobile– based story to meet another class requirement, or organize a workshop similar to the Up River workshop. These experiences can be used to test learning activities and build your and your students' capacity for doing ethnographic and mobile design work.

### 4. collaborate with students

At the Up River workshop, teachers and students learned how to use ARIS in teams. Several students picked up the software quickly and helped teach other students and teachers. In some cases, students' previous gaming and media design experiences came into play, as it helped them understand some of the language embedded in ARIS (e.g., quest structures, NPCs, items) and develop design ideas. Also, engage students as co– designers when you develop pilot projects, by actively seeking their advice for how to improve the project and embedded activities. These same students can serve as experts or classroom assistants when the project is implemented more broadly.

### 5. experiment with design

Build students' interest and expertise with the tools through game jams, where they build games that are not related to any specific content. This allows students to initially explore the tools and design processes, before adding the constraint of specific content or concepts. Later, students can use their understanding of the tools to design media that aligns more closely with your specific curriculum area.

### 6. look for content– rich places and issues

While all places are rich in content, we have found it useful to design mobile stories around places and issues that are already well documented with source materials readily accessible for classroom use. For example, one reason we selected Grassy Point as one of our locations was because we found a rich cache of photos, text, and video associated with this place.

### 7. teach ethnographic skills

Fieldwork is critical in any place– based learning. The practices we described above for immersing ourselves in and documenting the estuary need to be developed by teaching fieldwork skills. Observation, for example, requires more than "pay attention" and "take notes" — students need examples of very specific things to pay attention to. Students need more than a list of questions to engage in successful interviews — they need demonstrations and practice in class of how to follow up on answers, so that listening is emphasized more than asking questions. Students also need practice in class with using cameras and digital recorders — even a short lesson in class can improve the quality of video and photos. Although observing places and interviewing experts are important for studying ecological environments, these techniques should be supplemented by monitoring with digital probes. Finally, field data must be analyzed and discussed before students understand it well enough to select data to represent the place they are studying.

### 8. consider multiple uses for media

If you are doing a project around a particular place or issue, a mobile media game might be the only one of the things students design. They could also use their ethnographic work to publish a book, organize a photo exhibit, or start a website where others can add their own stories. Also consider contributing the media you create to the public domain (e.g., through Creative Commons) so that others can use it in the future. Use release forms to obtain permission from people early in the process, since it is easier to get permission when you initially interview or photograph someone than to secure it later. Finally, develop a system for archiving fieldwork documents and media for use in future projects.

### *9. cultivate partnerships*

Identify and recruit partners who can help with all aspects of the project. Our partners in Up River were invaluable. They shared content and resources, helped us understand key concepts, provided feedback on our design, and helped us connect with an audience of local teachers and students. Partners can make classroom projects richer for students and help teachers design more complex place– based learning experiences.

## CODA

While acknowledging the constraints associated with integrating new content, technologies, and practices into the curriculum, we believe that engaging students in ethnographic fieldwork and design presents unique place– based learning opportunities. The complexity and richness involved in designing ethnographic mobile stories increases students' motivation, provides wonderful opportunities to learn new skills and content, and develops students' under-standing of local human and ecological systems. Perhaps most importantly, it encourages students and teachers to ask new questions about where they live and fosters a connection to place. We hope our own adventure inspires you to explore your local community. We'll see you upriver!

# launching investigations with bite–sized gaming

*by Bob Coulter*

# launching investigations with bite–sized gaming

## by Bob Coulter
*Litzsinger Road Ecology Center,*
*Missouri Botanical Garden*

A flock of fifth graders stepped off the bus at the ecology center, ready to start their field investigation. They had studied ecosystems in their classroom, and learned all the right terminology: producers, consumers, decomposers, and the like. Armed with notebooks, they had been well prepared by their teacher to make the most of their time in the field. There were notes to be taken, measurements to record, and comparisons to be made. One element would be different, though: Students would spend part of their morning working in teams with an augmented reality game called "Who Rules the Forest."

Teams of students equipped with handheld computers explored the site, meeting different virtual game characters inspired by the Missouri bottomland forest ecosystem. Each character made an impassioned defense of why they are the most important part of the ecosystem. Clarence Cottonwood went for the size angle, arguing that he was the "biggest and baddest" part of the forest. Bella Beetle on the other hand went for the functional argument, noting that without her and other decomposers the forest floor would be littered with dead plants and animals. Harriet Hawk saw her usefulness in managing rodent population levels. Each species made a balanced, scientifically valid point about their significance in the overall ecosystem. At the end of the game, each team was challenged to identify the most important member and justify their selection. As they did this, they based their argument on what they learned from the virtual characters, the field observations they were prompted to make during the game play, and whatever other knowledge they brought to bear on the task. As you can imagine, this provoked some lively discussions as students debated different points of view.

In the end most students came to the expected conclusion that all parts of the ecosystem were interdependent. A few of the more creative thinkers pushed the game further, arguing that in fact Stewart Soil was most important. The others could be replaced by another of its type (e.g. Clarence Cottonwood by Ollie Oak), but that without soil, the rest of the ecosystem would collapse. In short, nothing else could replace the functional role of the soil. Experienced teachers know this is not uncommon: In a productive learning environment students often surprise us with the depth and clarity of their thinking.

This game is one of several developed by the Missouri Botanical Garden in partnership with the Scheller Teacher Education program at MIT to promote environmental investigations in a fun and challenging way. Other examples of projects we've developed include a watershed study conducted in a local park and a scavenger hunt within a cemetery. For the watershed project, students started the game by meeting a stream ecologist who challenged the students to investigate a water quality issue in the park. While the specific investigation was fictional, it was based loosely on an actual issue that the city and the Environmental Protection Agency were working to resolve. The challenge here was to decide — based on the clues within the game — which of several possible sources was causing the pollution. The game mechanism guided the students to observe actual pollution sources within the park to help them decide which were significant and which aligned with the clues to the mystery posed at the beginning of the game. The net benefit to the student game players was a greater awareness of their neighborhood park as an ecological space, and an enhanced understanding of key water quality issues. This experience in turn provided a foundation for subsequent water monitoring undertaken by the students in their after– school science club.

Lest we think these projects are only useful in the sciences, an early prototype we developed supported students in a 99% African–American neighborhood as they came to understand why they were practically surrounded by Jewish cemeteries. A scavenger hunt type of game had players "meeting" the people who were buried there, reading what history could be gleaned by the game designer from Census records, and in many cases being "introduced" to relatives and neighbors of the deceased who were buried elsewhere in the cemetery. Through the game experience, students gained an increased sense of what their community used to be.

While each of these games only lasted 30–45 minutes, their educational value endured much longer. Beyond the enthusiasm that you would expect from students playing a game with new and "cool" technology, the real benefit was in the continuing discussions sparked among the players. The relatively short time investment in the game play laid the foundation for rich, longer–term investigations. For example, the ecological studies that followed "Who Rules the Forest" were enriched by the students' understanding of interdependence; the group playing the watershed game embarked on a long–term monitoring and stewardship project with a greater awareness of the park as part of a larger ecosystem; and the grave–walkers started a study of local history imbued with a sense of those who came before them. In each, the game proved to be a catalyst for something much deeper than a typical introductory activity. While your neighborhood will have slightly different opportunities, the same type of experience is easy to create.

## DESIGN GOALS

In our experience, when you get kids outdoors exploring in an authentic problem–centered environment they will take their thinking past textbook orthodoxies and see things more creatively and more thoughtfully. The mobile augmented reality platform provides a vitally important resource for higher–order thinking, as it motivates students to observe more closely and integrate different elements of their understanding. This chapter explores ways to use augmented reality in the context of very short games designed to introduce new fields of study. While we have yet to try it, augmented reality (AR) games could also be used as a capstone near the end of a study. There, the challenge might be to integrate and apply what has been learned to solve a more complex mystery.

In either case, these are games a teacher can generate in a couple of hours and deploy on the school grounds or in a local park. In an age of prescribed, locked– down curriculum, this bite– sized entry to AR gaming can be an appealing place to start. With a minimal investment of classroom time, students can have a meaningful experience that carries through into other parts of the curriculum.

Now, on to specific design goals:

### promoting place– based education

As a conceptual framework, place– based education (Sobel, 2004) takes students away from the anonymous anytime / anywhere framing of knowledge found in most commercial curriculum packages. Instead, students are rooted in the place they know best: their local community. Botany is explored through plants in the schoolyard, park, and nature preserve; history through the town's past. From this launching point, students can encounter the world. How are the plants in the rainforest different from ours? Why? Or, How did our community change with the coming of the railroad? This anchoring in the immediately surrounding area doesn't work for all curriculum areas, of course, but where it is possible the local frame provides a solid base from which to work. In our experience and that of others, this connection to the local makes the learning more meaningful, and it has the potential to spark a greater stewardship concern. People are much more likely to act to preserve what is important to them.

### building environmental literacy

Paired with this emphasis on the local is the premise that solid, age– appropriate environmental literacy should be a pervasive part of the game design. In the examples given earlier, the forest game embeds accurate characterization of the different denizens of the forest, along with higher– order concepts such as prey– predator relationships, food webs, and interdependence. The watershed game embeds examples of both point and non– point source pollution, and examples of the types of testing that indicate water quality problems. In most cases, these same tests are ones that the students will be doing as they take responsibility for monitoring the stream in the coming weeks. The art form here is to include science at a level just sophisticated enough to engage and challenge the kids, extending but not overtaxing their growing understanding of environmental concepts. Textbooks are notorious for being jargon– laden. There is no virtue in simply turning a game into an electronic dispenser of poorly understood terminology. Rather, a good game environment provides the story, characters, and artifacts around which new scientific terms make sense.

### personalizing and anchoring

Looping back to the place– based curriculum frame, anchoring the game play in the local space helps provide a context for students as they wrestle with environmental concepts. For example, the nitrate level in the stream becomes more important to students when it its their stream that's affected. This all comes together as students make the links between recent rains, the seasonal fertilization of the golf course up stream, and the nitrate levels found in the water. Taken together, the creative fusion of a place– based approach to education and a focus on developing environmental literacy shifts students away from memorizing the textbook and toward building a strong conceptual network.

### using gaming 'hooks'

Underneath these meta– level goals, there are specific game design techniques embedded in these projects. First, in order to fulfill their role of sparking interest and understanding, the games need to have lively and creative "hooks" such as playful characters and a compelling story line. We knew that the players we were designing for had experienced science as a very academic endeavor, with an undue focus on reciting formal terminology as a sign of understanding. To lighten the mood a bit, we designed the game with a playful twist, embedding fun characters such as Bella Beetle and Susie Squirrel into the game. Designing for other audiences might suggest other important considerations. What is important to note here is that the characters didn't become caricatures. While the game play was fun, it was grounded in authentic, age– appropriate environmental science.

### keeping a limited scope

Along with a good hook, we felt it was important to maintain a light touch and not try to "cover" the entire curriculum scope in the short duration of the game. We wanted the kids to understand the big picture and to have just enough of a conceptual structure that they could build on as they continued their study. Even if being standards— based is an important design consideration, it doesn't mean that all standards need to be covered that day. As teachers, we need to resist the pressure to make the games all things to all people. Good design is often a process of subtraction more than it is of addition.

### watching our language

Another design goal was to literally "watch our language." As noted above, there is a danger in science education of becoming terminology– driven, infusing our curriculum prematurely with the official language of science. While we want the kids over time to learn to speak the language, the design process needs to take into account the students' current level of comprehension. In one case, we found this language gap to be a real problem, as we took the forest game that was initially designed for highly literate fifth graders off the shelf to use with a more mixed– ability group of fourth graders. While they managed to get through the game, it was clear that they hadn't absorbed nearly as much of the subtly embedded environmental concepts. Instead, they spent more time decoding the language on screen. It's also likely that they had less outdoor experience and a weaker overall science curriculum, so concepts that the fifth graders from one school assimilated well were rather fuzzy for the fourth graders from another school. Taken together, it reminded us that a game designed for one audience may not be appropriate for another, even if the age difference isn't all that much.

## PRACTICAL CONSIDERATIONS

For the teacher or environmental educator looking to design games such as these, there are a few other considerations we have found to be worth considering:

### *Picking a Good Location*

The first consideration is — to steal a line from real estate — is "location, location, location." The three examples described at the beginning of the chapter each leveraged what was available in the specific locations where the game was to be played. The forest game took place in the bottomland forest of our ecology center; the watershed game in a local park with a stream running through it; and the cemetery game in an historic local cemetery. On the other hand, some project ideas just don't lend themselves to the space available. One school, for example, made a valiant effort to help students engage in tornado readiness through an augmented reality game. Unfortunately, there aren't many locations on a school yard that have clear relevance to understanding tornadoes. The end result was a game that had students move from place to place on the school grounds, but the specific locations weren't at all relevant to building understanding. They were simply directed to different parts of the parking lot to learn something new before moving to another spot. The game was just a fact– gathering event that was functionally independent of the space in which it was played. The key here is to be a good location scout and use what is available. Don't force the game where it doesn't fit. There may be a better way to engage students than through a mobile game.

### *Time and Context*

Another consideration is to be aware (and plan for) where the game fits in the larger educational context. The forest game was planned to be just one part of the morning that students were spending at our ecology center. The time they spent at our site in turn was just one component of an ecology unit that would include classroom work, explorations in the woods near their school, and return visits to the ecology center. Thus, the game was designed as one component of a fairly intensive program in which the students would be participating daily. On the other hand, the watershed game was designed for an after– school environmental club that would be meeting weekly, with the inherently variable attendance pattern that after– school programs tend to experience. Given this difference, each game had to be designed somewhat differently. The game targeting the more intensive study could have more "hooks" that would be revisited over the course of the ecology unit, while the watershed game had to have a more narrowly constrained range in order to keep focus on the most important program goals.

### *Play Testing*

Once the game is designed, it's time to go out into the field. The first rule here is one that I learned in my professional internship many years ago: Always do a dry run yourself before trying a project with the kids. For one of our first AR ventures I was running behind schedule and tried a game "live" with kids that hadn't been play tested. We quickly learned that a setting in the game kept the students' play time to ten minutes, half of which was actually burned up waiting for the GPS signals to lock on to the handheld units. Imagine our collective frustration when the games timed out, with no real recourse other than to go back to school and re– load a fixed game file on to the handhelds. If this wasn't bad enough, it happened on an unusually cold and windy day. Needless to say, it wasn't the best start we could have had. Blue– lipped students rarely enjoy what they are doing.

Even if there are no overt design problems like these, there are a number of potential challenges to address when you are "live." One common problem is the inherent variability of GPS signals and the accuracy of the aerial photography upon which the game is based. It's best to test ahead of time to see what safety considerations might come into play. Your carefully positioned creekside target may well put the kids right in the water. Quite often, the accuracy issues can't be overcome, and you need to use some design kludges to focus the player's attention. In the woodland game, we learned pretty quickly that our intended locations rarely triggered exactly where we wanted them to. Bella Beetle may be at work decomposing a log in the woods, but she would "trigger" on the students' screens perhaps 15 feet away. Even on the same day, she would be in one place for the first group of players and another place for the next group. Woodland projects seem to have particular problems with this, as the tree canopy interferes with the GPS signals. Given this, we took to using small utility flags to mark the intended locations. Pragmatic, but not perfect. The cemetery game worked out a bit better, in that we could "pin" the locations to known grave markers, each showing a picture of the intended target on the screen. So, if students are looking for a specific name on a headstone, the marker is built into the landscape.

## IMPLEMENTATION

On the day you are unveiling your game, resist the urge for a major introduction. Give the kids just enough of a hook into the project to get them started. For example, you could tell them that they are being asked to solve an environmental mystery, and that they will pick up clues along the way. Explain that the mobile media is there to guide them. Then give them just enough direction to operate the units and get out of the way. Let them figure out how to navigate— the icons on screen will tell them if they are getting closer or farther away from their target. After the first stop or two, they will be confident navigators.

If anything, you will have to slow them down to notice relevant clues at each site. We have found it helpful to have students work in pairs, alternating roles as the navigator (with the handheld) and recorder (noting clues and other data on a sheet with a clipboard). This shared effort sparks discussion about what is being observed, and it helps to have this dialogue going when students are called on to make strategic decisions. As the game moves forward, students should be encouraged to use what is in the game and in the field to add to their knowledge store, working toward resolving the puzzle or dilemma that frames the game.

The game itself, of course, should always have a clear ending point. Even if the specific answers offered by each team may vary, all of the kids should know when they are done and what the next steps are. Often, this next step is to reconvene as a group and use what has been learned to answer the puzzle that started the whole quest. If so, perhaps the last stop in the game can remind the students of the initial question and encourage them to review their notes in preparation for the group discussion. While the game itself is over at this point, your teaching role has just begun. Your skills as a discussion leader will be paramount.

Before moving on, one final note on implementation: We always make it a point to tell the players never to let the handheld guide them to unsafe places, such as out into traffic or into a creek. We also try to have an adult leader such as a classroom teacher or one of our volunteers at least within eyesight of each group, unobtrusive but available to help as needed. While we want to promote independence, we also need to manage real and perceived safety concerns. In many neighborhoods, turning young students loose with expensive smart phones can make them crime targets. More generally, parents and schools today are much less comfortable turning students loose to explore the neighborhood than they used to be. You'll need to address these concerns in a way that is right for your kids and your community.

## FEEDBACK

As you might expect, virtually all students enjoy playing these games. The risk here is accepting the positive feedback too easily as a sign of the program's success. As the saying goes, any day on a field trip is better than a day in the classroom. Throw in some new smartphone technology and you have an instant winner. Generally, kids will latch on to new technologies with a great deal of enthusiasm. The challenge for us as designers and educators, though, is to filter students' responses to see what the real educational value is. For that, we need to look at the ways students engage with the problem at hand, and not just bask in their enthusiasm as a sign of how talented we are as game designers.

At a basic level, we want to know whether students are connecting more deeply with their community through the game play. Do they observe more carefully? Are they thoughtful about what they see? In the watershed game, we noticed that students who lived in the neighborhood and played in the park quite often still discovered new things as a result of the game play. For example, the dog walk area at the edge of the park went from being peripher- ally noticed at best to being a central player in the unfolding environmental mystery. We also need to be attentive to what players do with the information they gather. Are they processing deeply, or just rushing to complete the game? For example, there is an inherent tendency among kids to like the cute puppies at the dog park, but we need to note whether the kids think analytically about the impact of the dogs and their waste. It was rewarding to hear a student discount the dog walk as a possible cause of the pollution problem, correctly noting that "the doggy doo thing is too far away from the stream."

Since the premise of these brief games is to be a catalyst to subsequent learning, we need to be watching for how the gameplay feeds into later experi- ences. A number of anecdotal observations show increased understanding and awareness of community issues in later work. For instance, students who are out testing water quality have been more attentive to what is upstream from their testing sites instead of just taking the test results "as is," and our budding forest ecologists show more systems– level thinking as they analyze food webs and prey– predator relationships. When there is a hawk overhead, we've seen nurturers in a group want to protect the songbirds, while other kids were ready for the hawk to make a kill. Both responses show that the kids "get" the interactions unfolding in front of them. Back in the classroom, it is not uncommon for students who had played the forest game to pepper later discussions with how a change would impact the game characters. For example, one girl noted that "if there were fewer rodents, Harriet Hawk would go hungry and die." All of this provides good feedback to us as game designers, suggesting that even brief AR gaming experiences can have a powerful impact on students' interest and disposition to learn.

## TAKE– AWAYS

Looking back on our work with these bite– sized AR games, we are pleased with the progress made to date. We're confident that students can gain a great deal from a fairly short experience, which speaks well to the viability of AR gaming to be a useful enhancement to a time– crunched curriculum. Based on this early success, we see a great deal of potential for others to adopt AR gaming as part of the regular school day, in after– school settings, or as part of a field experience offered by a local cultural institution.

The technical demands aren't huge: Each iteration of the augmented reality software lowers the bar in terms of the technical skills needed by the game designer. This allows greater focus on the most important element— designing the students' experience. Where do you want the students to go? What should they experience while they are there? How can you talk through the virtual characters to the student game players, guiding their perception and building conceptual structures? What will students do with what they learn through the game environment? This is where the real teaching and learning takes place.

Mobile learning through augmented reality is a powerful tool, but it is not an end unto itself. Using new technology doesn't make one a 21st century learner – only an educator can. But it does help to have the right tools for the job. AR can be one of those tools — I invite you to try it out.

### reference

Sobel, D. (2004). Place– based education: Connecting classrooms and communities. Great Barrington, MA: Orion.

76

# beetles, beasties and bunnies: ubiquitous games for biology

*by Louisa Rosenheck*

# beetles, beasties and bunnies: ubiquitous games for biology

*by **Louisa Rosenheck***
*MIT Scheller Teacher Education Program*

## UBIQBIO SNAPSHOTS

Thalia is a 9th grader on her way home after school and volleyball practice. She's tired after a long day and not looking forward to the pile of boring home-work she is supposed to do, so she's thinking about skipping it. She gets on the bus, sits down, and takes out her smartphone. She puts in her earbuds and turns on her current favorite music. She sends a few texts and checks Facebook. Then she switches to the web browser and navigates to a game called Invasion of the Beasties. In this game she has to genetically engineer a monster so it will be able to defeat the evil enemies. Thalia likes this game because it has quirky characters with funny pictures, and it's fun to advance through the levels. Her teacher has assigned everyone in her intro biology class to play the game, but Thalia doesn't think it feels like school. It's obviously about mRNA — there are nucleotides and amino acids — but it's like a puzzle you have to figure out and she wants to see if she can beat the game before her friends do. Plus she's always on her phone anyway, and she can fit this in on the bus and between classes. Thalia has played five rounds when it's almost time to get off the bus, so she puts her phone away, knowing that the game has saved her progress so she can pick up where she left off later tonight.

Ms. Geary is a high school biology teacher finishing up some grading and getting ready to head home as well. The quizzes she's grading are tedious so for a quick break she opens up the teacher web site for Invasion of the Beas-ties, which she has assigned her students to play. She checks to see who has the highest score today, and while the usual students who reliably do their homework are indeed in the top five, she's pleasantly surprised to also see Thalia listed there. Ms. Geary has always seen Thalia as a perfectly capable student, but it's been hard to get her engaged in any of the biology content she's taught this year. Looking at the stats, she is happy to see that Thalia has been playing about 30 minutes each day this week, and considering the high level she has reached, Ms. Geary knows that Thalia is understanding the topic of mRNA and the translation process. If she wasn't able to use the Universal Genetic Code, she wouldn't have earned such a high score in the game, so the data is very promising. Ms. Geary closes the page and reminds herself to congratulate Thalia in the morning and make an effort to encourage Thalia to share the strategies she's developing with other students. While these mobile biology games are fun and engaging for many students, one of the most valuable things Ms. Geary has found is the way they also stimulate content– rich, authentic conversations.

## WHAT ARE UBIQGAMES?

Invasion of the Beasties is just one game in the MIT Scheller Teacher Education Program's series of Ubiquitous Games, or UbiqGames. A genre of mobile, casual, educational games developed at MIT, UbiqGames have a number of unique characteristics pertaining to the games' format, content, and feedback mechanism. In this chapter I'll start by describing the design principles of the UbiqGames genre, then give four examples of Ubiquitous Games designed for biology. Next, I'll explain the UbiqGames approach to how this style of game can be implemented in schools and support existing curriculum. Finally, I'll present feedback from students and teachers who experienced these games as part of their intro biology course.

UbiqGames are designed for the small screens of smartphones, but they are browser– based games rather than apps, so they can also be played in any browser on a desktop, laptop, or tablet. In addition, player logins enable game data and progress to be stored on a server. UbiqGames are casual games, meaning they are quick to learn and can be played in short amounts of time — with play sessions lasting 5– 10 minutes. This can work well for learning games since students can play in frequent, short bursts, during interstitial moments of their day, without taking up precious class time.

The games are closely tied to the curriculum and focused on one topic. Additionally, they are often simulation– based, giving students the opportunity to play around with the content and explore at their own pace, as often as they need to. This allows students to engage more deeply with the material at the same time they are exposed to the concepts in class.

The Teacher Portal is a web site teachers use to track student progress. The data– logging system collects player information which is then displayed to teachers so they can easily track participation. Along with game data such as score and level, the games also log other types of data that describe play patterns such as how long players spend in each area of the game, and how many times they log in. In addition, teachers can collect data that will reflect how well a student understands the material and therefore help teachers assess their progress. These data are displayed in the Teacher Portal to let teachers quickly see how much their students are playing, and what they are getting out of their play time.

From their unique format, content, and feedback capabilities, we can see that UbiqGames and the accompanying Teacher Portal are designed to be engaging for players and facilitate deep learning, while providing a feasible way for teachers to integrate them into their curricula and lesson plans. UbiqGames can be a valuable tool for teachers who are looking for a flexible way to incorporate mobile games into the often constrained classroom environment.

## UBIQ DESIGN GOALS

The UbiqGames approach was designed to address some common challenges of teaching high school science, for example:

| Challenge | UbiqGames Feature |
|---|---|
| Engagement: Students are not engaged with biology topics. | Good games are naturally engaging and motivating, and can highlight the intrinsically interesting aspects of biology |
| Pacing: Students don't all learn in the same way or at the same pace. | Simulation— based games let students "mess around" as much as they need to and make mistakes or progress at their own pace. |
| Exposure to content: Students stop thinking about typical lab activities or traditional homework once it's been completed. | Students play UbiqGames more frequently throughout the week, increasing the number of times they are in contact with the material. |
| Limited school— based technology: Teachers can't always schedule time in the computer lab or with the laptop cart; there isn't enough time to give up class periods for online activities. | UbiqGames utilize out—of—class time and can be played on a variety of devices, anytime and anywhere. |
| Limited student— based technology: Students may have limited computer access both at school and home. | More and more students have their own phones and even their own smartphones with internet access which they can use to play UbiqGames. |
| Feedback opportunities: It's hard for teachers and students alike to monitor their progress. | The Teacher Portal collects relevant data and displays it for both teachers and students to quickly see how they're doing. |

## THE UBIQBIO PROJECT

Our biggest UbiqGames project to date has been the Ubiquitous Games for Biology (UbiqBio) research project, funded by the NIH from 2009– 2011. We worked closely with biology teacher consultants to design and pilot the games and create supporting curriculum to help facilitate transfer. We then worked with a group of Boston– area high school teachers to implement the games in their classrooms. During each relevant curricular unit in the semester, teachers introduced the UbiqBio game in class and loaned their students Android smartphones provided by MIT. Teachers chose to assign the game as home-work (or in some cases for extra credit), and students played outside of class over the course of the week– long unit.

Periodically throughout that unit and especially at the end of it, teachers used class time to tie the game content back into their curriculum. They drew on examples from the game and facilitated discussion around in– game metaphors and strategies, thereby encouraging transfer from the gameplay to a broader context. Specific ways of doing this which teachers found most effective will be presented in the Curriculum Design section of this chapter.

During the implementation of the games, MIT researchers collected a variety of data including content assessments, surveys, observations, interviews, and user– generated log data. One goal was to analyze this data to learn more about the efficacy, engagement, and feasibility of the UbiqBio games and approach. Another goal was to learn more about what types of casual games would best engage, motivate, and teach students. By studying these aspects, we were able to paint a clearer picture of the successes and challenges of UbiqGames for biology while also learning more about best practices for designing educational games in this genre.

### a suite of games

Each of the four UbiqBio games covered specific standards— aligned curriculum points in the areas of Mendelian genetics, mRNA and translation, evolution, and food webs. We worked with teacher consultants to identify these as areas with which students often struggled. Each game also had a unique game mechanic, premise, and visual style, resulting in a diverse suite of games that balanced cooperative and competitive play, realistic and fantastical worlds, and simulation– and narrative– based games. An overview of each game is presented here with some discussion of the most interesting features of each.

## BEETLE BREEDERS: MENDELIAN GENETICS

*Premise: Customers want to buy beetles with certain traits and it's your job to breed them! Choose the contracts you want to work on, then mate the right beetles to produce the desired offspring. Use your knowledge of Mendelian genetics to work with increasingly difficult patterns of inheritance and maximize your profits. How much money can you earn in the beetle business?*

Beetle Breeders facilitates learning for students by allowing them to play through just one or two contracts per session, and still get the experience of trying different crosses and seeing the results right away. One thing that sets the Beetle Breeders game apart from the others in this series is the richness of the biology content. A basic Punnett square mechanic lets students experi-

ment with a wide variety of inheritance patterns, which makes it relevant to a larger chunk of the genetics unit. Players work their way up through the levels of the game encountering more complex tasks as they go. This keeps the game interesting and makes it easy to see their progress as they move through the curriculum.

## INVASION OF THE BEASTIES: mRNA & TRANSLATION

*Premise: Strange and scary monsters are taking over! You must genetically engineer your own band of monsters so they will be suited to fight each opponent. Use the Universal Monster Genetic Code to research which proteins you need to synthesize. Adjust the nucleotides in the RNA strands and match the correct amino acids to create polypeptide chains without mutations. If you're successful, the resulting phenotype will give your monster the ability to defeat the enemies!*

The biology content in Invasion of the Beasties is more specific, focusing on the concept of the Universal Genetic Code and the relationship between nucleotides, amino acids, and phenotypes. There are three levels of increasing difficulty, but since the game covers fewer biology concepts, "beating the game" feels more attainable. More than our other games, this one is very narrative—

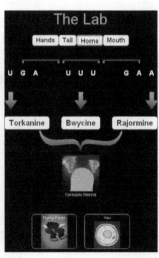
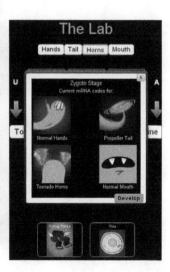

based and character– driven. The illustrations bring the game to life and players enjoy giving each other tips on how to solve the puzzle of which phenotype will defeat each enemy. The highly simplified models of biological systems (such as amino acids, the universal genetic code, and genetic engineering) require teachers to highlight and explain the differences between the game and real biology, but they also make the content more accessible and fun for the students.

## ISLAND HOPPERS: EVOLUTION

*Premise: In a world full of islands each with their own bunny population, small changes to the environment can have noticeable evolutionary effects. You have the power to make environmental changes on your own island, such as increasing temperature, adjusting the local flora, and even introducing a virus. By collecting data over many generations and looking at the proportion of certain traits in your population, you will discover evolutionary trends and learn to predict future population changes.*

Island Hoppers poses an interesting design challenge because our goal is to take a very broad topic, evolution, which often contains many misconceptions for students, and break it down into bite– size chunks. We ultimately do this

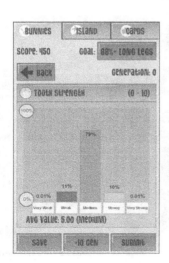
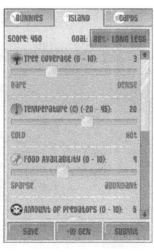

by specifying formulas for the back– end simulation which enable players to fast– forward through time step by step to research each relationship. One element unique to Island Hoppers is the graphically represented data and the use of histograms to display the breakdown of certain varieties of bunny traits. Reading graphs is a very important skill in science which is sometimes overlooked, so teachers value this feature. At the same time, the graphs provide information on whether players are making the right moves, so students are motivated to learn to read them.

## CHOMP!: FOOD WEBS

*Premise: Mysterious species are connected in complex food webs that are under attack. Aliens have been chomping on these ecosystems and each time they decimate one species, it has a drastic effect on the other interconnected species. Players must examine the relationships between species to understand and predict the population increases and decreases. If they can use this knowledge to determine which species was the latest victim, they will be able to restore the food web to its balanced state!*

louisa rosenheck

The main goal of Chomp! is to give students practice reading the energy flow of a food web. It presents just two modes of difficulty, offering relatively less

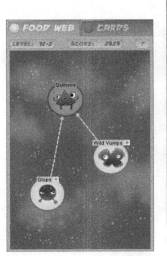

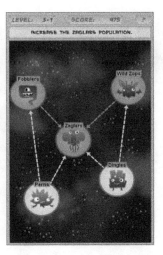

gameplay than the other games. Like Invasion of the Beasties, Chomp! utilizes fictional species content, which in this case compels players to think about the predator— prey relationships rather than relying on prior knowledge of real species. Because each food web is generated procedurally, no two players get the same sequence of puzzles, though the generated food webs do increase in complexity. As a result, instead of students' sharing the specific solutions they find for each puzzle, they are encouraged to explain the overarching concepts in order to help each other.

## CURRICULUM DESIGN

UbiqBio games ideally should be fun as standalone games, but we don't expect students to learn everything they need from the game on its own. On the contrary, the design of the implementation and surrounding curriculum are equally important. UbiqBio games are designed to be implemented in conjunction with the relevant unit in a biology course and teacher facilitation is essential to their success. Teachers can use UbiqBio games to introduce a concept, to explore it more deeply, or as additional practice or review. Here is a typical implementation example of using a game in the middle of a unit to let students gain experience with the ideas:

**day 1:** Introduce genetics, genotypes, phenotypes, and Punnett squares.

**day 2:** Demo Beetle Breeders game and assign students to play 20 minutes a day for the rest of the week.

**days 3 and 4:** Continue teaching various inheritance patterns as students use out– of– class time to explore these concepts in the game.

**day 5:** Debrief the game with students, discussing their breeding strategies and how the game is like or unlike real genetics.

Because transfer is one of the most challenging aspects of any educational game, we worked with teachers to design curricular materials that would address the issue of transfer explicitly. For instance, given the implementation above, these are some of the possible ways to connect the game content to the class content:

**warm–ups:** Problems that get students thinking in the same way they will need to in the game, such as determining the possible phenotypes of a child with two blue–eyed parents.

**do–nows:** Tasks with a format very close to what students need to do in the game, such as selecting parents for a cross or completing a Punnett square.

**worksheets:** Activities that extend the premise and characters of the game, such as working backwards to identify the parents of a given offspring.

**in–class examples:** Ideas students come up with using the game as a frame of reference, such as identifying the number of spots beetles have as an example of a polygenic trait.

**discussion questions:** Questions that challenge students to compare strategies they used in the game, such as: How did you choose which beetles to mate?

**think–abouts:** Prompts that encourage students to think about how the game concepts relate to their own lives, such as: What traits have been passed down in your own family?

Both concrete materials as well as guidelines for best practices are an important part of the UbiqBio approach, and were refined throughout the implementation of the project. We relied on teachers to explicitly connect the games to their curriculum, frequently tying the game mechanics and content back to the unit once all students had that shared experience.

## UBIQBIO IN SCHOOLS

As mentioned previously, these games have already been used in high schools as part of a research study. During the spring semester of 2011, high school students at three Boston– area schools used all four UbiqBio games as part of their intro biology course. We worked with six teachers with a total of about 200 participating students. The grant provided smartphones with data plans (but no voice or text capability) so that each student could access the games anytime, anywhere, and in frequent but short bursts, as they were designed to be played.

At each point in the semester when the class came to one of the relevant units, the teachers would sign out a phone to each student and assign the game to be played as homework, usually over the course of about a week. Teachers attended a professional development session for each game and curriculum materials were provided as described above, but each adapted those to better fit into their existing curriculum and preferred activities. During the time students were playing each game, teachers could access the Teacher Portal to monitor student progress and get a sense of how they were doing. Available data included how often and how long students were playing, and how far they had progressed through the content of the game. Teachers either assigned students to play for a certain length of time each day, or to play through a certain level, both of which they could confirm through the web site.

For the most part teachers were able to use the games according to plan, though there were some challenges. Many of the difficulties were technology– related: phones malfunctioning or students forgetting to charge them, games running slowly on 3G networks or bottlenecks when many students played simultaneously, and school– based internet going down when teachers had planned to demo a game in class. These issues are inevitable in any educa- tional technology initiative, and certainly had an effect on our goals of giving students easy access to the games.

In addition, students are already using their own phones for many things in their lives — texting, music, Facebook, etc. — and being able to do schoolwork on the same device makes the transition more seamless, as opposed to using an "extra" device as they did in this case. We expect that a few years from now when more students have internet– enabled mobile devices of their own, many of these issues will disappear.

Finally, one of the main departures from our ideal UbiqBio scenario was that teachers ended up spending more class time on the games than they had hoped. It often took longer than planned to hand out phones, demo the game, and make sure everyone knew what to do, sometimes taking as much as a full class period. While these are essential activities, designing more tutorials and instructions into future versions of the games would move more of the start– up time from in– class to outside– of– class and provide students with the tools to figure things out on their own. This would simultaneously enable teachers to dedicate valuable class time toward deeper discussions of the game content.

Despite some of the challenges, teachers and students alike were excited to be using mobile biology games as part of their classes, and overall they enjoyed the experience. Through the variety of research methods that we used, we were able to get a good sense of what factors contributed most to the UbiqBio experience.

## THE STUDENT EXPERIENCE

Of the students that played the UbiqBio games, 77% thought they were fun, for a variety of reasons. There was definitely a "cool factor" of getting to use a new smartphone, but students also found the games fun and even addicting and said that playing the games did not feel like schoolwork. As one student put it,

*"It like gives you excitement in what you're doing. Homework, no you just sit*

*there and you're like 'No, I don't want to do this.' You want to burn the paper. But no, the game, you're just like 'Wow, I want to keep going and going.'"*

This type of response indicates that we were able to meet our goal of designing games that were fun and engaging. In particular, students especially liked Invasion of the Beasties because of its wacky characters and humorous art style, which suggests that narrative is an important part of mobile game design.

Even though students were assigned to play the games, they also enjoyed doing it and often played a good deal more than they were required. Sometimes students would sit down to play in one long stretch, and they most often played at home, but they also took advantage of the ability to play one round at a time while waiting for the bus, riding in the car, or between classes. Only 18% of UbiqBio students thought it was not easy to find time to play, and interviews revealed that the mobility of the games did contribute to students' willingness to use them since it was easy to play in any room of the house, or wherever they happened to be. Naturally no single learning tool will address the needs of all students, but with the majority finding these games fun and easy to fit into their busy lives, this approach is likely to be welcomed by many teachers.

Competition played a significant role and was a motivating factor in the play experience for many of the classes. Some teachers sparked the competition by using the Teacher Portal to announce the current leaders each day and even offering extra credit to the winner. However, students also got into the competitive spirit on their own, asking their friends what level they were on to assess the competition and staying up late some nights in order to have the highest score by the next morning. One student explained,

*"We felt like we were competing because people come in and say 'Oh, I'm in this level.' and then people are like, 'You are?' and then it's like 'I could beat that, I could beat that.' so it gives you motivation to go beat them."*

Although we didn't include a built– in leader board or direct in– game competition, these are clearly things that many students would be enthusiastic about for future UbiqGames.

We did find examples of students who were typically struggling or disengaged with class but then became engrossed with certain games. A few of them even surprised the whole class by earning the highest rankings. Their teachers were always thrilled to see that happen, and we feel that it demonstrates the ability of games to appeal to different types of learners and the value in providing opportunities to try things out on their own in a safe space.

Students' main reason for liking UbiqBio was that the games were fun, but 76% of them also felt that the games helped them learn. They recognized the fact that playing the games was more active than doing worksheets and they thought being able to really see things happen and make changes to the beetles, amino acids, etc. by themselves could help them understand the ideas. For example, after playing Beetle Breeders, one student said:

> *"With the Punnett Squares I think that helped me … to understand it. … I've taken this class before and I haven't gotten it, that's not something I'm good at. … It made me feel better because instead of having to sit in class and take notes there was a better way to learn how to do that. … I feel like now I get it more, I understand it more."*

Students said they would have liked to see more graphics and more action in the games, but our simulation– based designs were definitely a good start.

There were elements of certain games, such as Punnett squares and food web diagrams, which reminded students of their experience in the games when they appeared on tests and may have helped them transfer knowledge to these more traditional assessments. However, we also noticed that students saw the games as largely separate from class. Based on interviews with students, there appeared to be very few times when students were playing the game and consciously thought about something from class, or vice– versa. This suggests that future games and curricula might be improved by emphasizing connec- tions between in– class and out– of– class activities, and facilitating transfer between formats.

## THE TEACHER EXPERIENCE

Teachers also found UbiqBio interesting and useful. They had fun playing the games themselves, and all 6 teachers reported that their students were engaged with the games. They felt it really made sense to have activities that kids could do on devices they already know and love, especially ones that were so easily integrated into their existing lesson plans.

The Teacher Portal proved to be an invaluable tool for teachers, and despite not being as polished or user– friendly as we the designers had hoped, five out of six teachers said they did not find it at all challenging to use. One teacher described her use of the web site:

> *"I looked at the teacher portal … at least once a day. … If I noticed there was a student that had … very low points … then I knew that I needed to catch up with that student and find out what was wrong. … And so it just gave me some more details of where my students were at."*

Using the site, teachers could identify who might need more encouragement or help with the game, as well as who deserved recognition and might be able to give that help.

At the end of the semester, all of the teachers reported that they felt the UbiqBio games helped students learn biology content and practice related skills. There were a number of ways that teachers could tell UbiqBio was helping their students learn. Some felt that the games provided more back- ground for students so that when they covered the concepts in class, teachers could use game content as examples and the students could already identify with that. They also noticed that certain students had gained confidence using the concepts on their own in the game. One teacher, describing the in– class discussions, said,

> *"The kids that played [the games] that normally wouldn't be doing a lot of work, were much more willing to give their opinions. They answered a lot of questions and answered them well."*

Collaboration among students was one effect of UbiqBio that genuinely surprised a number of the teachers. One of them said,

> *"I was surprised about how much the students helped each other out. Many times students that didn't get it were offered help by another student in the class that did. I didn't think the games would generate that much cooperation."*

In this way, the games started conversations about biology content that wouldn't normally occur and teachers were thrilled to see that.

## TRYING THE UBIQ APPROACH

Teachers considering using the UbiqGames approach in their classrooms, whether with these or other mobile games, will want to think through some key aspects of the curriculum design:

- What do you want students to gain from their individual gameplay time?
- How does that experience relate to your existing lesson plans?
- How will you make connections between in– and out– of— class formats
- and concepts?
- How will you encourage student collaboration and discussion?

After studying the ways that UbiqBio games were used in six teachers' classrooms, we know that while each game had its pros and cons, these are some of the guiding questions which are key to a successful implementation. When thoughtfully integrated into a curriculum, we have seen that a successful UbiqGame can be a powerful tool for learning in any content area.

They engage and motivate students with a wide array of learning styles and background knowledge. Like other good educational games, they help students understand concepts more deeply and practice skills at their own pace. With proper support they are easily integrated into existing lesson plans and leave in− class time for the most valuable activities. Lastly, being mobile, connected, and personalized, they provide students with educational experiences that fit the way they live their lives.

## NOTE

At the time of writing, our four UbiqBio games are still under development, being cleaned up before being released to the public. Once the games are stable and the log data is reliable, the games and Teacher Portal will be available for free on the MIT Scheller Teacher Education Program web site. Interested teachers will be able to create an account, manage their classes, and access the games and related curriculum materials. For the most current information, please visit education.mit.edu or email *louisa@mit.edu.*

## THANKS

It takes a lot to design, develop, and research a suite of mobile games, and we would like to acknowledge everyone who played a major role on the UbiqBio team. Thanks to Professor Eric Klopfer, researcher Judy Perry, developers Susanna Schroeder, Fidel Sosa, Jose Soto, and Grafton Daniels, artist Amanda Clarke, and teachers Amanda Tsoi, Lauren Poussard, Lisa Curtin, Rebecca Veilleux, Leo Medina, and Emma Lichtenstein.

◉

# mystery trip: reframing with a narrative

*by John Martin*

# mystery trip:
# reframing with a narrative

## by John Martin

*University of Wisconsin— Madison, Academic Technology*

In this GPS– assisted game, campers embark on a "typical" 4– day trip. But once they're far across the lake, their reason for hiking changes! Through their communicator — their only connection to base camp — they hear terrible news from camp. With periodic updates, they figure out what's happening, and play the lead role in saving the camp (while learning much more than usual about the land they're hiking through).

**duration:** 4 Days
**location:** Wilderness of Maine

The Mystery Trip is an outdoor interactive adventure that increases engagement with "and connection to" place, helps build interpersonal relationships, and helps increase outdoor acumen. It was based on a trip sent out in the 1920s– 1940s that required too much set up time to sustain, but was updated and run using MIT's Outdoor AR platform on Windows Mobile devices and Bluetooth GPS units, between 2005– 2009.

Flying Moose Lodge is a deep woods summer camp in Maine that allows 10– 16 year– old boys to go out on four– day trips each week, where they canoe or hike in small groups. Extended outdoor adventures typically provide opportunities to learn a number of things in a context that is neither contrived nor artificial. The setting is authentic and the consequences are most– often natural, rather than imposed by a constructed rubric or educational plan. In addition to increasing technical skills such as canoeing or fire– building, partici-pants learn things that are hard to teach in a structured setting, such as the ability to improvise (e.g. how do we set up an effective campsite in these condi-tions), to think fast (e.g. it's about to rain, do we continue on route X, or find another?), to live with decisions (e.g. we should have stayed with our original route because this one is less protected), to accept with a certain level of grace that one cannot control all the factors one has to face (e.g. weather, terrain, group dynamics), and to appreciate the control and opportunities that one is afforded (e.g. attitude, skill– building, planning, a sunny break from rain).

Since 1993, I have helped direct the camp, and with other staff had noticed that while canoe trips taught campers to canoe well enough because every stroke of the paddle required concentration on the skill, they weren't learning many skills on hiking trips. Trails were, in the words of one camper, "like interstate highways" — there was no real need to think about what one was doing until the exit (trail junction). Consequently, campers talked about sports, girls, school, video games, etc. rather than about the experience they were engaged in, or about the camp community they were operating as part of. Decades ago, when campers would stay for eight weeks at a time, good outdoor skills, tight communities, and a deep sense of belonging would eventually emerge. As camp sessions became abbreviated because of other summer commitments, we also noticed that campers connected less — both with each other and with environment. We looked to emerging GPS technology, and MIT's Outdoor AR to recreate an old trip that hadn't been run since the late 1940s. A revitalized Mystery Trip provided an opportunity to address these issues by immersing the campers in an intense group– narrativized adventure. The old Mystery Trip, as described by a camper who participated in it in 1927, appeared to be something that mobile technology could support and enhance:

> *Towards the end of each summer, while the older boys were doing manly things on the Allagash or at Katahdin, we others took part in the wild pursuit of thieves, kidnappers, and other nefarious individuals.*

*That first summer of mine, quite unexpectedly, as we were about to set out on our regularly scheduled trips one Tuesday morning, we were all called together and the cold facts were put before us. Something terrible had happened; I am sure that I don't remember what. Plans had to be changed at the last moment, and all our energies were to be devoted to helping the local authorities, whoever they were, hunt down the criminals and bring them to justice. At the same time we would uphold the honor of the camp, and in all probability bring fame and fortune to ourselves and our counselors.*

*Assignments were quickly made. For the sake of expediency, the original trip groupings would be maintained, but we would travel unexpected paths. All of this had been well arranged beforehand; and I can visualize the counselors now constructing the complicated plot in the evenings after we had gone to bed. Now they were ready to play it out.*

*I can't remember much of that first Mystery Trip except that it rained. It rained all the time. The villains, whoever they were, had left clues and trails as they chal-lenged us to track them down. Coded messages were found and deciphered. The net was slowly tightening. In tracking those undesirables, we learned more than we at the moment wanted to know about following trails in the woods. I clearly remember looking for stone cairns on the mountain side under what were certainly not the most favorable conditions.*

This was too fun to let it remain in the archives — even with the rain. Present day campers were recruited to create a modern narrative equivalent, and came up with a story about the camp being taken over by a rival camp. It's a simple linear narrative with 13 parts/locations.

**1. start of trip:** Campers embark on a 4– day "Trails" trip, where they're told they'll be hiking some mountains in the 5000 acre wilderness next to the camp. I give them a "Communicator" (Windows Mobile device) and explain that we'll be trying to keep in touch with them in case anything happens. They canoe across the lake and get to the base of the first mountain when something happens.

**2. across the lake:** As they hike up the mountain on the trail, the communicator buzzes. It's the assistant director, looking beat up and not responding to any of their questions. He suggests that the communicator might have been damaged in the melee, but he hopes they can hear him. He has terrible news. The camp was invaded. He doesn't know by whom. It was chaotic. He managed to escape, and is spying on the invaders, so he'll keep them informed, but they need to get out of there because the invaders know about their trip. Go to the secret campsite on the top of the mountain and wait for more information, but stay off the main trails and be sure that no one sees them!

**3. top of mountain:** At the top of the mountain the communicator buzzes again. The invaders are setting up a radio tower. We can use the communicator as a receiver, and if we triangulate the signal maybe we can decode it and figure out how to defeat them. Get to the top of three more mountains to collect the signal. But camp here for the night.

In parts 4– 13, they make their way around swamps, through brambles and briers, up (and down) three more mountains to capture and triangulate the signal being sent out by the invaders. This triangulation provides enough information to decode the signal, which is then modified and rebroadcast in such a way to send the invaders away. Suffice to say, there are a number of huge holes and logical leaps in the story that always motivated good– natured debates and theories within the groups to explain what was actually happening in the narrative, and how triangulation could or could not provide decoding clues, etc. While our first instinct was to try to close the logical fallacies, we decided that if it encouraged the formation of arguments and peer– to– peer learning, it would be foolish to take them out.

To capture and triangulate the signals, and in the spirit of moving the narrative forward, the campers inevitably take more difficult and less– used paths – sometimes bushwhacking. They operate in stealth mode and employ Leave– No– Trace principles. Instead of talking about girls and classes and baseball, they talk about what might be happening in the narrative, learn map– reading and way– finding, and help each other through difficult terrain.

Essentially, during the four days they hike up the same mountains that they had planned to hike, and do all the same things they would have done without the interactive narrative. But the addition of the interactive narrative dramatically changes their attitudes, motivation, and participation in the trip. They feel central to the trip and to the camp because it's up to them to save it.

> <u>Camper 3</u>: *On a regular trip you just want to get to the next campsite, but on this you have to get to this mountain to stop the radio signals...*
> <u>Camper 4</u>: *And you don't know what's going to happen next! That's really fun!*

## DESIGNING MYSTERY TRIP

In designing a wilderness trip, our first concern is safety. In addition to the standard practice of keeping campers in groups of 3–6, supervised at all times by licensed trip leaders, we employed additional safeguards. Recognizing that they may stray from established trails, we sent each group with an additional GPS unit. We thoroughly briefed the trip leader on both the terrain they'd be traveling through, and the narrative that would structure their trip and kept the whole experience within 4 miles of the camp.

Beyond safety, I wanted to motivate and engage campers to participate in activities. Primary in our design goals is the need to make the design compelling. Through active participation comes competency and expertise in both interpersonal and outdoor skills, feelings of belonging and community, and a better understanding of and respect for nature. These are the desired outcomes of any camp activity, but they depend on active engagement by campers. If a camper does not want to be involved in an activity, he will not get as much out of it. We believe that nature offers plenty of problems to engage in and solve provided that campers have a reason to solve them.

Engagement was leveraged at the beginning of the design process by recruiting campers to make the game; to design something that would be used by their peers and potentially coming generations of campers. We sent the first group out with a notebook, GPS, and a description of the original game, and asked them to come up with their own story. They returned with a largely empty notebook (writing was too much like school), and some vague ideas and character descriptions, which we molded into a game (Wild Moose) primarily through discussion.

The next round of campers were then able to play that game. They were asked to play it and modify it based on their own ideas. They initially weren't impressed with the game the previous group had created, but having a concrete example of a game to play was enough to motivate them to "do better" and think through ideas more thoroughly. They had a complete narrative (Mitchville: Where It All Began) with characters and locations, which we were able to plug into the game editor the next week.

In this design, we used a very simple linear narrative and stretched it out over 16 square miles. The narrative just offered a framework for them on which to build their own team's adventure. As they played they continued to ask, "What happens next?" This approach to design and writing was enough to motivate the campers to crisscross over the mountains, through forests, and around swamps as they role– played a rag– tag group of spies.

## IMPLEMENTATION OF MYSTERY TRIP

Over the course of three summers, about 40 campers and staff "played" through the Mystery Trip in groups of between 3 and 6 people. Each group prepped for a basic hiking trip, and as they were headed out, I asked them to carry a "communicator" that we were trying out to try to keep in touch with trip groups. Most realized there was a game in action after the first communication from me, but there was a kernel of doubt throughout in some of the campers. Since the game narrative suggested that the camp was being taken over by a rival camp, and participants could see the camp from a distance, but only barely make out that some movement was occurring there (movement which was simply routine weekly maintenance), they could not really be sure what was happening.

Camper 1: *"I liked the espionage. Actually that was really fun. ... we have no evidence to actually disprove the game, so because of that, it was kind of fun."*

Camper 2: *"you actually have imagination because it says things that are going on that aren't really going on, so it's just kind of neat like that. So you can imagine what's happening in the game instead of just hiking."*

When they returned to camp I interviewed each camper individually or in groups about their experience and what they would recommend changing. Each group built on the ideas of the previous group, and redesigns became part of the process itself. Each implementation had challenges, surprises, and lessons for other iterations.

The challenges of running this game in 2005, at a wilderness camp that didn't even have electricity were numerous. We charged the devices in camp vans; software updates required driving into town twenty miles away; the Windows Mobile devices and Bluetooth GPS units were fragile and sometimes did not survive swims ("I forgot it was in my pocket"); and my one day off was typically spent writing up notes and tweaking the narrative based on participant suggestions. But the most frustrating thing was the battery life of the devices, further shortened on a number of trips due to campers playing Bubble– breaker and other games at night after hiking. In the morning, the "Communicator" has a dead battery. We found that having the counselor hold the device at the end of the day, and sending a second fully– charged device alleviated the battery problems. There were no game– related injuries in any trips..

## THE PLAYER EXPERIENCE

The sheer scope of a game to structure a four– day hiking trip allows a lot of breathing room in the game. In our experience, the simple narrative was allowed and enhanced by the mobile device, but was not centered in the device. We had no riddles to solve, but instead let the lay of the land between points provide the challenge. Similarly, we had no ecological lessons or morals embedded in the game, but instead let their own experience hiking and camping together supply the space and time to reflect on the lessons and morals that emerged.

The most surprising result was that the simple phrase "they're coming after you!" almost always prompted campers to bush– whack to their next destination.

> Camper 7: *"You think that you don't want to go on the trails because the other camp would be there waiting for you..."*

Given the choice, all groups reported going off trail, while the trip leaders maintained that they continued to practice many Leave– No– Trace principles. By choosing to go off– trail, they greatly increased the difficulty and skills needed to complete their trip. Off trail, the trees are closer together, terrain is uneven, and the direction of travel is uncertain. On a trail, there is a general set of expectations: that it leads where you expect it to; that it's recognizable and somewhat passable; that others have gone this way and survived. After a few trail– hiking experiences campers start knowing what to expect, and it begins to become routine; for many of the campers, hiking trips at Flying Moose had become that, as indicated in terminology used in the interviews, such as "just hiking" (camper 2), and "regular trail" (counselor 1). The "long snaky group" that occurred on regular trips segregated fast hikers from slow ones; the ones in front felt no need to help the slow ones because the trail was the obvious path, and more importantly they did not feel like part of a group on a group quest. By going off trail, they took out the map more often and gained orientation skills. They talked more among themselves about the best path, helped each other up cliffs, over downed trees, around swamps, and worked together as a team more than they did on other hiking trips.

> Camper 1: *"...and jumping from rock to rock at times, in fact there was one really deep hole with ground at the bottom that didn't look too sturdy. I'm like "Whoa! Don't fall down that" Of course J was really tired and was following me, and I didn't want him to fall in it."*

In some ways, the increased difficulty also increased the worth of the trip. As one trip leader explained, "you're not going to orienteer around a swamp, you're going to walk until you can put your foot down without sinking, and then you've got to figure out where you are again ... because of the sense of discovery that goes along with uncertainty, the rewards are in some ways greater than even hiking up a much more objectively spectacular mountain."

As they hiked, players engaged in building their own group story off each others' ideas, and new narrative elements emerged. Thus, the deer– hunting stands they stumbled upon, though not in the original narrative, became enemy sniper towers that must be avoided. And other people on the trail were hidden from because they might be enemy scouts. Because locations were set but routes were not, campers reported feeling like they were the first to step on that ground.

## LESSONS FROM MYSTERY TRIP

Through the narrative the campers saw themselves at times as fugitives, explorers, rescuers, and finally as heroes — more deeply connected to both the camp community and the land they had just experienced. The adventure was an initiation of sorts, and in "saving the camp" they were let in on a number of inside jokes that older campers and staff knew. Having now learned them, they too felt an increase in identity as part of the camp group. Furthermore, having experienced and co– created their own specific group adventure within the structure of the more generic narrative structure, they bonded with their trip mates, but also shared parts of it with campers who had not been on the trip — careful to include hints, but not spoilers, in case the others were going next time.

Their counselors reported that whereas typically campers dislike the difficulty of hiking off– trail, and whine and try to avoid it, in this case the narrative element gave them a good enough reason to try it, and having experienced it for the sake of the plot line, they learned to be adept and even comfortable with it. The difficulty of it, though, also decreased the relative difficulty of the other parts of the trips. For example, the tasks of setting up a campsite, building a fire, and making a meal seem luxurious after fighting ones' way through brambles. On regular trail–hikes, these things are a bother.

Surprisingly, map and compass skills also increased. We had thought these skills would decrease because the narrative was revealed by GPS on a Windows Mobile device. But the 200x200 pixel size of the screen just did not give enough detail or context to satisfy the campers' questions of how to get from point to point. They reported tremendous group discussion and map– and– compass work to correlate location and direction to landmarks and geographical features. As a larger object, the map was both easier to crowd around, and to contextualize their exact position relative to the land around them.

Though we had originally planned cut– scenes and a higher production value, we found that they were not needed. Much like a good book, the text narrative was enough to trigger the imaginations of the campers, and as they hiked they shared with each other what they thought must be going on at camp. And, it turns out that even the best computer effects cannot rival the resolution of real rain, the actual dropping of temperature at night, or the feeling of spider– webs smearing across one's face in a hike.

The hardware that we ran the Mystery Trip on is dead; the batteries no longer charge, and new Windows Mobile devices probably are not worth trying to find. The idea is a good one though, and it provided us with both a service-

able testing platform, and results that continue to inform. If we were to recreate the Mystery Trip again, we would (in 2012) probably simply port it into a platform like ARIS, and use iPod Touches and a portable Wi– fi hotspot that would be under counselor control.

One addition that we would like to add is to provide greater opportunities for participants to modify and add their own voices and experiences to the narrative. All found "hidden gems" in the various paths that they took. Many mapped these out on paper or on the emergency GPS, and some of these gems found their way into the official narrative. We'd like to make that process much easier, so that instead of playing one official version, campers could follow intertwining narrative threads laid down by those who had come before them — and leave some virtual threads of their own for those who come after. For example, the ability even to map out wild raspberries, blueberries, and strawberries could further motivate future campers to undertake the grueling bushwhack to get there, and forever immortalize the discoverers' adventure; saving a group from inadvertently wandering into a swamp, or pointing out a particularly scenic outcropping are all contributions that imbue a sense of philanthropy in the campers who add them. Sometimes giving back to a community is the best way to increase ones' sense of belonging to the community. We would like to provide that option when we re– create the Mystery Trip.

## SUMMARY

Augmented Reality authoring platforms, such as ARIS, or the MIT Outdoor AR platform used in the Mystery Trip allow communities to create place– based experiences. These can re– frame one's experience of space *in situ*, and foster deeper connections to communities and the cultures of place. This type of learning links community involvement with meaningful practice in physical place by tapping into deeply embodied pedagogies of sensation and cultural practice.

# mentira: prototyping language–based locative gameplay

*by Chris Holden and Julie Sykes*

# mentira: prototyping language–based locative gameplay

## by Chris Holden and Julie Sykes
*University of New Mexico — Albequerque*

A relative has just been accused of being involved in a murder and you are responsible for clearing the family name. You quickly learn that you will have to find clues by exploring a Spanish– speaking neighborhood in Albuquerque and sharing these clues with others. You think, "Spanish?! How in the world am I going to do this in Spanish? Why would I want to?". Yet you forge ahead into the adventure and begin to unveil the mystery of Mentira — a place– based mobile game designed for your Spanish 202 class. Over the course of the next four weeks, you play the game at home and explore a local Spanish– speaking neighborhood while, at the same time, improving your abilities in Spanish.

In this chapter, we unpack the Mentira experience in terms of design, implementation and evaluation. We first explore Mentira as an example of the potential uses of place– based mobile games. This is followed by a discussion of design goals, lesson learned, and future considerations. Our hope is to not only provide a case that examines place– based mobile games for language learning, but also give insight into the design and implementation of place– based games in other disciplines.

## THE MENTIRA ADVENTURE

*Mentira (http://mentira.org)*, which means 'a lie' in Spanish, is an augmented reality murder mystery game that is integrated in a fourth semester Spanish course at the University of New Mexico. The unit takes place over four weeks. The game is set in a Spanish– speaking neighborhood in Albuquerque, Los Greigos, which eventually becomes the site for a field trip in which players use local resources to access in– game content, and, ultimately, solve the mystery of Mentira. Students play the game on iPod Touches or iPhones which are either owned by the students themselves or loaned to the students as part of the course. The game is made accessible via iTunes for those who wish to use their own device. Levels 1– 1, 1– 2, and 2– 1 prepare players to go to the Los Griegos neighborhood by introducing them to their family, revealing the murder, and expressing the need to find clues in order to absolve their own family and preserve the honor of the family name.

As the game begins, the player joins one of four families which each have different family values, expectations for behavior, and insider knowledge. For example, the Silva family is a long standing law enforcement family with a great deal of respect in the local community. They value integrity and are brave, fair, and a little rebellious. In later iterations of the game, these characteristics are first presented to a player on a family card in order to set the context and establish some parameters for interaction (see Figure 1).

FIGURE 1 — FAMILY CARD: SILVA

A player's family determines the type of information he or she is given, some personality– based expectations for interaction, and sets the context in which the rest of the mystery is eventually revealed. Different families have different clues and provide information that is needed by all to solve the mystery, creating an information gap scenario in which players from different families must collaborate in order to get all of the information. For example, a member of the Gurulé family is the only person to get a specific clue about a notebook found near the scene of the murder. Therefore, as the game progresses, players must work together and share information in order to have all of the clues and be able to solve the murder. The next two levels of the game take place over the next two weeks and are focused on setting the scene and preparing the player to go to the actual neighborhood to find clues.

## DESIGN GOALS

A number of design goals centered on enhancing the language learning experience of students through the use of a place– based AR game. Here, we highlight a few that are especially notable.

### *leverage place*

It seems logical that Albuquerque, NM would be a great place for an English speaker to learn Spanish. After all, Spanish was used in the community long before English and continues to be an integral part of much of the city. New Mexico was part of Spain and Mexico before it was part of the US. Yet, in all reality, the majority of Spanish classrooms at the University of New Mexico look much like any other Spanish classroom in the country. In many cases, our students feel that what they do in the classroom is abstractly and distantly related to their lives. Mentira is an explicit attempt to change this, to make the local Spanish– speaking community a meaningful part of learners' language experiences. Therefore, the game comes out of Albuquerque's identity as a place bound up with the Spanish language and makes concrete use of the rich context immediately outside of the classroom. It is intended as a stepping stone for students into the Spanish speaking world, not another tool for memorization and regurgitation of language forms.

Visiting the neighborhood to play the game is an important part of the authenticity we are trying to impart, and part of the reason Mentira is in the form of a mobile game. Technically, this story could be told in a desktop game, or entirely in the home environment. People could play it anywhere in the world. But without going to the neighborhood, the experience falls flat. We will discuss later how we see more curiosity, willingness to take risks, and teamwork out in the field than in the classroom with Mentira. The coordination between the players' physical locations and the place in the game world is the center of the magic that is augmented reality. Generally, its what makes this new genre interesting.

### an interesting story

To get students excited to play Mentira, we sought to create a story that:

- did not feel fake and arbitrary
- was not a tourist perspective
- connected not just the language and a place in time and space, but also felt like it belonged to the rich culture that surrounds Spanish in the local community, and
- was exciting enough to make for a fun game.

FIGURE 2. LOS GREIGOS, 1883. EDDIE ROSS COBB. COBB MEMORIAL PHOTOGRAPHY COLLECTION, UNIVERSITY OF NEW MEXICO. CENTER FOR SOUTHWEST RESEARCH.

Center for Southwest Research, University of New Mexico

FIGURE 3. PLAYING MENTIRA IN MODERN LOS GREIGOS

The story, coupled with authentic language, provided a game that was realistic and relevant for the place. We decided on the format of a murder mystery, historical fiction, because it allowed us to create an authentic setting from actual historical details and gave us the freedom to create a simple and direct goal for the game as a whole — solve the murder. We found a nearby neighborhood, Los Griegos, that used to be an agricultural village before it was swallowed by the growing metropolis of Albuquerque. The village is at least as old as its 300 year old church, and Spanish has always been a part of the place. Traces of the past (see Figure 2) are everywhere to be found, and Spanish is still a part of Los Griegos' present (see Figure 3).

We also chose this neighborhood as a setting for our game because we had contacts who knew and had written about the history of the place. This gave us details that we could use together with more general themes and histories around Albuquerque to construct our narrative.

In fact, the name of the game itself comes from an interaction with a local resident about the neighborhood's history. In our documentation of the neighborhood, it was suggested that a particular building was a boarding house for priests or nuns. A neighbor we met in the neighborhood went across the street to ask her mother, a long– term resident of Los Griegos, about this. When she returned, it was with a single word "mentira" — a lie! Her mother strongly doubted the scholarship in our hands and sparked a name that now forms the foundation of the rest of the story. The interweaving of past and present, makes a strong impact on our plans for the future and highlights the way in which augmented reality can not only help students access the place, but also shape their interactions with it in the future.

### in Spanish, not about Spanish

One of the biggest shortcomings in educational materials for language learning is the way they position the learner in relation to the language. Rather than providing opportunities to use the language, they often focus on memorizing things about it. This is especially the case for programs designed for mobile devices. In an effort to deepen students' experience, Mentira does not try to teach about Spanish. It is in Spanish. Students use Spanish to play it.

With authenticity in mind, Mentira is designed to include a fair amount of vocabulary that the students are unfamiliar with. The reason is to reinforce the authenticity of the game as a story coming from a particular place, as dialogue with particular people. Rather than speaking generic textbook Spanish, these characters have regionally appropriate dialogue and their personality comes across in their patterns of speech. So we expect students will have to learn some new words in the course of the game, either by looking them up or asking others.

Much of the game consists of simulated dialogue with characters in the game. Our dialogue is very similar to the dialogue found in Where in the World is Carmen San Diego or Mass Effect. To interact, the player reads text coming from the virtual character and then usually has a series of choices of how to respond to that character. Depending on the choices made, this dialogue proceeds further and in different directions. FIGURE 4 is the beginning of an interaction with a member of the Silva family.

As you are introduced to Alejandro Silva in the first level of the game, he suggests you come back the next day to learn more details. You are then given a variety of choices about how to respond — walk away, give a rude response, and give a polite response. Each choice results in a different outcome ranging from game over to additional information. As the player interacts with more characters, language becomes more and more vital to continued success. Different personalities prefer different language and different strategies.

Success and failure in Mentira does not depend directly on the mastery of a word list. The unfamiliar vocabulary is part of the setting. There is not so much of it that students are totally lost, but enough that they know there is something missing. Instead of revolving around the assimilation of vocabulary, the conversations work in terms of pragmatics: knowing the social setting and acting appropriately. Characters in the game will tell you more or less, and sometimes outright lie to you, depending on the way in which you converse with them. For example,

FIGURE 4. TALKING WITH ALEJANDRO SILVA

FIGURE 5. IGNACIA JARAMILLO

Ignacia (Nacha) Jaramillo inquires about Alejandro Silva, a long time ally of the Jaramillo family, before interacting with the player (FIGURE 5). If the player is rude or does not indicate he or she is a member of the Jaramillo family, no clues are given and advancement is not possible. Getting enough information to help solve the murder depends on the students' abilities to suss out certain basic features of Spanish pragmatics. Also, the humor in Mentira results from opportunities to flagrantly go against the rules.

Pragmatics is a very difficult aspect of learning a second language, and so most classrooms ignore it. There are no hard and fast rules and so it is hard to test. We don't claim that Mentira is the missing teaching tool for pragmatics, only that it returns the topic to a place of prominence, and again adds to the authenticity of the story we are telling. The game does not call attention to them by name, but since the conversations revolve around social niceties, pragmatics becomes an obvious topic of conversation around the gameplay, whether directed by the teacher in the classroom, or by students in discussing the mystery.

Though the dialogue plays a large role in the game, students make use of Spanish in more ways during its play. In the game software itself, there are written clues, media items, and directions in Spanish whose successful decoding allows players to navigate the mystery and eventually find the real killer. In addition, the classroom activities, and our jigsawing of the mystery across several player roles, not to mention the field trip, create opportunities for the students to interact with one another in relation to the language in the game.

### jigsaws, roles, and other gamey goodness

In designing Mentira, we looked to borrow other features from the world of games that we felt were responsible for meaningful play and could support interactions outside gameplay. For example, to encourage teamwork across students, the clues to the murder (or their veracity) was scattered across four player roles in a classic jigsaw format. No player has enough information to solve the murder on their own. The clues only make sense by triangulating them with those obtained by others. Besides scattering information in the software, many in– class interactions, including the final solution to the murder, were designed to take place through small and large group discussion. There are also situations during the field trip where players have different possible locations and interactions based upon their chosen family. Through this jigsaw design, we emphasize the important role of collaboration in the gameplay experience.

One of the more basic benefits we hoped to gain from using a game was to produce an environment where language production had lower perceived risks because the game could always be begun again with no penalty. To accentuate this, we created many dead– ends, situations the player could get into (usually an unforgivable offence to a primary non– player character) from which further gameplay was impossible. Unlike most school work, we wanted failure to become an accepted part of participation in Mentira and to be productive for learning.

### combine classroom, home, and field work

One of the unique features of Mentira as an AR game is the length of time over which students play, and the different contexts in which that play takes place. As we mention above, students play over the course of two to four weeks. We lend them mobile devices during this time. The first parts of the game occur in the classroom and at home, and the peak of the game is a field trip to Los Griegos. Finally the mystery is solved back in the classroom.

We strove for such a long time frame, and gameplay that cuts across these contexts, because there was an element of continuity missing from earlier AR game experiments. Even when there were long game curricula, the students would only have the devices and access to the game software for an hour or so. This limited time frame is very different from how people use these devices in their real lives. The device is an individualized, almost omnipresent part of your life; it's there when you need it and goes with you throughout your day. Implementations with small, tightly regulated windows of access do not represent the potential of this technology for our students. To avoid this problem, we sought a way to integrate mobile devices into student learning in a more naturalistic fashion through the Mentira experience. Continuity of the device and software allows us to connect the classroom with what students are doing at home and more importantly with their experiences of the neighborhood where the game is set.

An important part of this cross– context design is the visit to the neighborhood. Certainly, taking students there follows through on our commitment to engagement with authentic place as it relates to the language. But the field trip also has consequences for an educational attitude we think is important to encourage in all learners — particularly so for those learning a second language — risk taking. This is where you step outside your own comfortable little world to experience something new. Even though Los Griegos is right in our backyard, most of our students do not have a familiarity with it or other historically important neighborhoods within the city. Forcing them to leave the classroom and go there is an important symbol of going out into the world and seeing something new, a critical assumption that underlies much of the promise of AR curricula generally.

### the student side

Getting feedback from players and incorporating it into the next iteration is an invaluable part of this design process. Mentira did not spring into the world fully formed, and the difference between the game and our vision of what it could be is truthfully pretty large. However, we have made a lot of progress over the last couple of years by listening carefully to our students. Their feedback has encouraged us to believe there is something interesting in Mentira, to continue work on this experiment. Beyond this basic assurance that we're barking up the right tree, feedback ensures our work is relevant to those who will be using the game in the future, and helps us understand some of the obstacles that lie ahead, some of them outside the realm of curriculum design.

An important aspect of using player feedback is deciding where to focus attention. It is easy to get bogged down in details that are not very relevant to the big picture. With Mentira, our main goal was to examine the potential of AR games to connect students meaningfully with their language learning experience. As a result, when students play the game, we pay particular attention to how they play and whether the game feels authentic to them. Part of what we hope to see in how students "play games" is for them to engage with the software and get into the story. More importantly, we were looking for things like impromptu collaboration, risk taking, role playing, learning to play vs. playing to learn, and taking ownership of their experiences within the game world. These practices are particularly productive for second language learning but seen more typically in videogame playing than in classrooms. To assess the authenticity or reality of the world we created for the students, we look for them to make connections between the game, learning Spanish, and the context of the Los Griegos neighborhood.

Student reactions to specific characters, the structure of dialogue within the game, or other smaller details of our implementation have been helpful too, but they are of lesser importance. The students' relative inability or unwilling-ness to "play" in the classroom, and the potential of the field trip to unlock this frozen state is the biggest lesson we have learned through having students play Mentira.

### classroom culture is robust

One purpose of Mentira, as described above, is an attempt to disrupt certain common classroom behaviors that are not productive for second language learning (e.g. risk aversion). And one thing we can see clearly in our players' actions is that the game itself is not enough. We designed a jigsaw to encourage students to collaborate, but when we observe classes after students begin play, actual collaboration between them is rare unless specifically asked for by the teacher. We gave them roles to inhabit, but their perception of the different personalities involved has been rather dim. We created a lot of failure points and gave ample space and time to allow for productive failure, but most students do not try again after reaching a failure state. Broadly, we want them to "play" our game, but, until they go into the field, they largely continue to see it as a homework assignment to be completed and graded.

Perhaps this is to be expected. Students who are not used to working collab- oratively are not going to fall naturally into it because we assure them that this one time it will not be cheating. We won't likely see these classrooms change from a teacher– directed, individual mindset to a student– centered, distributed one just because we put a game in front of them that responds to this kind of thinking.

These issues go deeper than the artifact. When someone like James Paul Gee points out that we see more productive learning behaviors around games than we often see in classrooms, this is really about changing the basis of social interactions that take place in classrooms, not about putting videogames in them. From this perspective, it seems like our software and instructions simply are not enough on their own to displace such a large situational reality. The good news is that another part of our game — the larger reality of the neighborhood where it was played — does seem to be strong enough to have an effect in this direction. By leveraging it more effectively in future versions of Mentira or in other projects, we believe we could do more to disrupt unproductive aspects of classroom culture and improve learning.

It would be unproductive to blame classroom culture for all the shortcomings of Mentira. We are careful to recognize the questions of execution worth addressing here as a brief aside. Just because designers make a role for players to choose doesn't mean that role is worth inhabiting. It also doesn't mean the interface for the game doesn't get in the way of the story. Execution matters, not just vision. When evaluating game curricula generally, one of the most common mistakes is to let the single example stand for the general case, to forget about execution. Game quality is especially important in AR games, a genre that is still being born. One of our hardest jobs as designers is to try and figure out which problems with our game will respond better to improved execution versus rethinking from the ground up, when our difficulties are about gaming and when they are about our game.

### the importance of place

The behaviors we wanted Mentira to provoke in students: playfulness, inventive-ness, collaboration, and risk taking — the behaviors that did not manifest in the classroom — emerged spontaneously in surprising ways during the field trip portions of the game. Students worked together as groups, helping each other with the conversations, vocabulary, and decisions about where to go and what to say. We even saw some groups decide to role play in the neighborhood — only speaking in Spanish for the entire duration.

In our interviews with students who played the game, we see that the reality and engagement of the game as a whole hinge on the robustness of their experience in Los Griegos. Here are some representative examples from players.

- "The first part before we went to Los Griegos, that was kinda boring, but when we were actually in Los Griegos and seeing all the clues and all the new information, that was a lot better."
- "It (going to Los Griegos) made it more real, and like um you weren't just like reading something out of a textbook or just doing some activity on a piece of paper, you were actually like walking around and it like put you in the place, and like you actually got to more apply your Spanish to like situations. Just made it more hands on."
- "It was different. It was a breath of fresh air. You know you could have had the game here at UNM, and it would have been kinda like okay, and you know kinda boring. Going somewhere else, having somewhere actually, have the game actually at a place set in around Los Griegos, um it was a difference, it was a breath of fresh air, and I enjoyed it, you know, immensely."

Even those for whom the experience fell flat saw a deep connection to place as something important and within reach:

- "Cause we were only there for half, you know a hour, we were only supposed to be there an hour or two so... Well I think, I think the idea of going out into communities, which is Spanish speaking neighborhoods and stuff is good. But it helps if there's someone on the street."

Our field trip is by no means a deep ethnographic investigation into the neighborhood, but it nevertheless has a large impact on the students who play Mentira. Going to the neighborhood sparks players imaginations for how they could get involved with real contexts in their quest to learn Spanish and how activities from the classroom might connect to those real world situations. The real Los Griegos makes our depiction of it in Mentira come alive for students.

### we don't wanna go

Despite the positive effect going to Los Griegos can have for our students, there has been substantial resistance on the part of the students to scheduling time outside class to visit Los Griegos to play the game. The hesitation doesn't seem to have anything to do with the location itself or playing the game, but rather the idea of committing to anything outside the scheduled hours of the class. Although younger students are eager to get outside the classroom, our students are coming from a different perspective: school is allotted a certain amount of control over their time, and the field trip is adding on top of that. This resistance is to be expected and may likely remain. The good news is that when students actually go on the field trip, they tend to view it as the best part of the game, value the educational experience we've created as a whole, and have a strong preference after playing for increasing the depth of interaction with the neighborhood.

## LESSONS LEARNED

In the sections above, we have tried to paint a picture of the promise we see in re– centering learning activities around local place using an AR game, and some of the main challenges we've faced in trying to make that happen for our students with Mentira. Going somewhere new can get students thinking outside the usual situation of the classroom, more broadly and meaningfully about what they're trying to learn. It establishes context for the rest of the game and curriculum. Beyond this basic lesson, there are a couple other take–aways we'd like to share about doing this kind of work.

### iterate, iterate, iterate

We cannot emphasize the importance of frequent iteration and testing enough. Long development cycles, complex creation tools, and complex media tend to get in the way of rapid iteration. If you have to wait several months for an art team to get assets to you before you try your ideas out, not only will you be stuck waiting when you could be testing your game, but when you need to make changes, your turn around time will be much longer. Use simple tools and focus on speed of iteration as the primary way to improve the eventual quality of your game.

With Mentira we made some decisions that have helped us iterate quickly. ARIS, our game engine, is very simple compared to other game platforms. Although this limits what our game can do — for example, we cannot use audio in our conversations, only text — we do not require the skills or time of a programming team to make changes. The limitations of a simple platform have another beneficial effect: fewer art assets to develop and maintain. If Mentira had audio, it would be much more cumbersome to produce changes to the dialogue in the game. Even having made these decisions, if there is one thing we would change about our development of Mentira, it would be to iterate our design more often.

In the end, it is not so much any one factor of production that tends to slow down the process of iteration. Instead, the biggest obstacle can be a desire for perfection. One of the lessons from videogames that we'd like our students to learn, and that we should mind as well is performance before competence. That's to say the efficient process to a good game is more likely to start out with a game — even a bad one — than to hope to make that game perfect

FIGURE 6. TESTING A VERY ROUGH VERSION OF MENTIRA.

before it ever reaches students' hands. Figure 6 shows our first test of Mentira with a class of students in the summer of 2009. The game was far from finished, and we had to use desktop emulators in a computer lab instead of mobile devices in Los Griegos. Yet the feedback on our vision and design was extremely valuable.

In the age of rigid assessments and a severe efficiency paradigm, it can be harder than ever to adopt an attitude of try– and– see, but it is not reasonable to do something that has never been done before and hope to know all the details in advance.

### keep your eye on the prize

In the section on student feedback above, we mentioned that we mostly try to focus on the big picture when considering changes to the details of Mentira. To extend that thought a little further, when thinking about next steps, we think how we can serve our ultimate goals — even if it means leaving this game behind or vastly changing what it is and how it is used.

For example, an unfinished area of Mentira is in how the mystery is solved through collaboration and the use of clues. It largely happens outside the software due to time constraints and the simplicity of the platform. With time for another iteration of Mentira and a newer, more capable version of ARIS, creating an in– software ending is tempting; it sounds like a way to vastly improve the game. Yet incorporating the solution of the murder in software may not be our first priority. Although doing so would increase the robustness of the game experience for students, we are not sure it would do the most to bring them closer to the Spanish speaking world that exists outside their classroom doors.

Instead, perhaps we will seek ways to integrate the game more tightly with the neighborhood it takes place in and references. We may even look for alternative means to create experiences that are designed to capitalize on the excitement and interest already created by the game. This follows from our criticism of typical classroom practices and artifacts as too narrowly focused on their efficient use, that they never get around to involving students with the worlds in which the language has meaning. From our perspective, it would be narrow minded to entirely limit our focus to improving the quality of our game if that focus does not lead to an improvement in our ability to get students into Spanish speaking worlds.

130

# place–based design
# for civic participation

*by James Mathews*
*Jeremiah Holden*

# place–based design for civic participation

## by James Mathews and Jeremiah Holden
University of Wisconsin — Madison

Over the past three years we have partnered with a group of teachers, educational researchers, and community members to collaboratively teach a place–based high school course called People, Places, and Stories (PPS). One of the key goals of PPS is to engage students in identifying and researching cultural and ecological themes and issues in their local community, then designing media and events (e.g., documentaries, photo exhibits, games, community events, and digital stories) to share their findings and personal perspectives on these issues. In recent implementations of PPS, mobile technologies have emerged as key tools for supporting students' fieldwork and shaping the media products and experiences they design throughout the class.

At the end of each semester of PPS, we interview students about their learning experiences. Here are some typical responses:

> "Nobody ever really pays attention to what goes on in their community, just themselves ... [this class is] kind of teaching you to actually pay attention."

> "We're playing [place— based] games and designing some games too. And I think that if other kids got to experience that, it would help them learn a little bit better because we learn while we are building the games. So we are putting it together and learning at the same time ... Instead of just reading it or just looking it up."

> "Well the game got other people's attention when it came to what was going on

*in the conservancy ... Maybe after they played the game they wanted to get involved or paid more attention to what was going on in the conservancy."*

*"It's kind of like we're making our own history not just learning about other's history."*

As evidenced through these quotes place– based mobile games and stories have become central media for engaging students in exploring and representing important people, places, and issues in their local community. It is important to note, however, that while we sometimes design mobile games and stories for our students to play, a central component of PPS involves students designing their own mobile media. To exemplify what this process looks like, we now present two projects that emphasize students as designers.

## PROJECT 1:
## THE NEIGHBORHOOD GAME DESIGN PROJECT

The Neighborhood Game Design Project (NGDP) was implemented with assistance from Mark Wagler, a folklorist, teacher, and educational researcher (Mathews, 2010). NGDP included three curricular components that unfolded over a three month period. The first was a series of place– based inquiry activities where students used mobile media to identify and investigate contested places and issues in their city. As part of these investigations students engaged in basic fieldwork activities (e.g., mapping, interviewing, photography) and generated questions about what they might investigate further. The second component included a series of mobile design workshops where students individually and collaboratively designed games and stories using mobile devices. Workshop goals included introducing basic design processes, learning the features of different mobile game and story platforms (e.g., Scvngr, MITAR, ARIS), and exploring how these tools might be used to engage others in thinking about the contested issues and places under investigation. The third component was a sustained design project where the entire class collaboratively researched a community issue and designed an Augmented Reality (AR) story to teach other students and community members about the issue. In this example students designed the AR— story To Pave or Not to Pave, organized a kick– off event to officially release their design, and facilitated a research activity that assessed changes in users' perspectives surrounding the debate.

### topic selection

After an initial brainstorming session aimed at generating potential topics, students decided they wanted to learn more about a recent proposal to redesign the local nature conservancy. A key feature of the proposal called for paving one of the main paths through the conservancy — an option many found disagreeable. Because it runs adjacent to their school, many students felt a sense of ownership over the path. Additionally, students believed the city was moving forward with the plan despite strong public opposition. Having identified a topic, students brainstormed possible design ideas, eventually settling on a place– based mobile story.

### place– based mobile design

Students decided that their final design should present multiple perspectives surrounding the proposed plan and encourage users to reflect on their own perspective about whether the path should be developed. The story would raise other students' awareness about the issue and contribute to the debate. It would end with players having an option to sign a petition asking the city to place a moratorium on future development in the conservancy. To achieve this vision, students quickly realized they needed to learn more about the development plans, identify key issues and perspectives surrounding the debate, conduct fieldwork and documentation in the conservancy, and learn more about mobile storytelling and the associated technologies.

As students moved forward on this project, their design– based learning occurred across four interconnecting components of PPS:

1. Fieldwork: Students conducted interviews and user surveys; engaged in mapping activities; and gathered video and photographic documentation at the conservancy.
2. Open Lab: Students researched relevant web– based and print resources; communicated with community members via email; created and organized media; built prototypes; and engaged in informal critique and feedback sessions.
3. Design Meetings: Students discussed and voted on core design decisions, reported on their progress, and made plans for next steps.
4. Playtest Sessions: Students presented and tested design prototypes, then discussed ideas for refinement.

### mobile– based interactive story

As a result of the mobile design workshops, students produced an AR place– based story called To Pave or Not to Pave. In this interactive story visitors to a local nature conservancy meet a "concerned citizen" who informs them about a city plan to pave a path in order to create an off– road transportation route. He then asks if they would like to sign a petition opposing the development, suggesting they learn more about the issue before doing so. As visitors walk the remaining sections of the path they watch videos and interact with virtual characters who share their perspectives on the development plans. These characters, all of whom are based on real people the students interviewed, also share their knowledge about the conservancy's history, ecology, and use. At the end of the path visitors learn it is too late to fight the development plan, but are asked to sign a petition asking city council members to pass a resolution restricting similar development in other parts of the conservancy. This real– world petition, which was created by students in the class, demonstrates two key points about the team's design choices. First, it shows students commitment to representing the debate from multiple perspectives instead of producing a story that was explicitly persuasive. Second, students hoped the story would lead to an immediate action, with visitors making an informed decision whether or not to sign the petition.

### research and release event

In addition to the AR design, students planned and hosted an event where other students at the school played To Pave or Not to Pave, then engaged in a discussion about whether or not the conservancy should be protected from future development. At this same event they used pre– and post– surveys and focus– group interviews to test the impact of the design on user's knowledge of the conservancy and their opinions related to future development. The students used this feedback to reflect on their design and make recommendations for how it could be improved.

## PROJECT 2:
## THE CAPITOL PROTESTS

The Capitol Protests originated from students' interest in studying a series of protests occurring in and around their state capitol building. The project was implemented with the help of Jeremiah Holden, a teacher educator, researcher, and former civics teacher. Like the Neighborhood Game Design Project, the Capitol Protests included three curricular components that occurred over four weeks. The first was a series of inquiry activities where students studied and discussed ongoing protests occurring at the capitol building in response to pending budget legislation. The second component included fieldwork whereby students visited the protests as citizen ethnographers and documented the events using a variety of media and methods. The third component was a series of Augmented Reality design workshops, where students first played Dow Day, a situated documentary about anti– Vietnam War protest in their city, then prototyped several similar AR designs aimed at representing some of the core perspectives, debates, actions, and experiences associated with the current protests. In describing these curricular components it is important to note they were not enacted in a linear progression. For example, documentation and design often generated new questions, leading in turn to additional inquiry discussions, online research, and fieldwork.

### topic selection and initial inquiry

When we held one of our initial meetings to decide what topics or issues students were interested in studying during the semester, the protests quickly emerged as a primary concern. Due to the local political actions and prominent national media attention surrounding the state budget legislation and negotiations, most students were quite familiar with the protests and a few had participated in them with friends and family. A few students, however, knew very little about the political issues or related civic activities. Regardless, the consensus was that the budget legislation and the protests were important events worthy of our attention. Additionally, the students felt they were uniquely poised — if not responsible — to document the events because they were happening then and now, in their backyard. PPS provided the infrastructure, scaffolding, and flexibility for students to act.

Having chosen to focus their research inquiry on the protests, students first constructed a timeline of events, identified key political issues and politicians, developed a set of inquiry questions, and compiled relevant web– based and print resources. The conversations and discussions that emerged during this inquiry stage were quite dynamic, in part because students were eager to share their own perspectives and experiences. With little prompting, the students asked clarifying questions, gathered additional background information, and identified key issues and questions they wanted to explore through fieldwork, documentation, and additional online research.

### fieldwork and documentation

A primary goal of fieldwork and documentation was for students to develop a better understanding of the state budget legislation and related political protests. To meet this goal students visited the state capitol as citizen ethnographers. In this role they were trained to conduct interviews with protesters, gather video and photographic documentation, and employ other field research methods. Engaging in ethnographic work at the Capitol provided a unique opportunity for the class to experience the protests from a different perspective. For example, one student who had protested at the capitol viewed this visit as an opportunity to interview people about their own stories related to the protests. For another student it was an opportunity to see with her own eyes what was happening without having to rely on the media or friends' stories. She particularly tuned into how the unique constraints of television shaped how the events were being reported. An important aspect of students' experiences as ethnographers was that it helped them generate new questions and encouraged them to experience the protests from multiple perspectives, both of which later informed their discussions, insights, and decisions as media designers.

### augmented reality design

Students' fieldwork and documentation set the stage for their follow– up design activities. After sharing media, themes, and questions that emerged from their fieldwork, students voted on whether or not they wanted to continue their inquiry and use their findings as the basis for designing a mobile– based game or interactive story. In the end, three students decided they wanted to do this for their final project for the semester, while others chose different questions and issues to pursue.

Like To Pave or Not to Pave, students' AR design work was scaffolded through a series of studio– based workshops. To initiate this process students began by playing Dow Day, an ARIS– based situated documentary about anti– Vietnam War protests which occurred in their city in 1967 (Mathews & Squire, 2009). Playing Dow Day introduced students to the functionality of ARIS and AR design, including increasing their familiarity with features such as quests, characters, and items. As a group, students first produced a simple mobile experience based upon their fieldwork. In this first iteration there was no narrative arc; rather, students created items and characters based on what they learned through their documentation as citizen ethnographers. For some students this encouraged design experimentation as they "translated" field– based interviews into ARIS– based characters and dialogue. It also provided an opportunity for them to reflect on and share their own perspectives on the protests. A second series of design activities were organized for students to conceptualize and prototype an ARIS– based mobile story aimed at teaching others about the protests. As a result, students designed a narrative– centric AR game where players had to survive a week at the protests. The students borrowed heavily from their own gaming experiences, as well as their experiences as protesters and ethnographers to produce a game that focused on key events, issues, people, and perspectives surrounding the protests.

## place– based mobile storytelling and civic participation

Three key themes relevant to student learning and civics education emerged across both projects. First, students exercised choice about what they learned, which cultivated a sense of ownership over the learning environment and media they designed. Second, aligning media design with local issues opened opportunities for students to share their voices and participate in public discourse. Third, perspective recognition, a core skill associated with democratic participation, emerged as an important part of students' inquiry and design work.

### choice and ownership

The design of PPS's classroom environment and curriculum, particularly its emphasis on democratic participation, encouraged choice and ownership to emerge as normative elements of teaching and learning. Not only were students involved in co– designing their learning experiences (e.g., by helping decide the places and issues we studied), they also had autonomy when determining the content and goals of their final designs. In order to support these ideals, we developed protocols to guide group decision– making and allow students to move between the classroom and the community as needed.

PPS's democratic practices provided opportunities for students to manage much of their own learning and pursue personal interests. Some focused upon a particular skill or media practice, such as video production or game design, while others were able to study a particular topic or issue they found meaningful. In one instance a student used PPS as an opportunity to further explore his interest in photography. Another student's work with the conservancy project aligned with an interest in birds; she used the AR design as an opportunity to learn more about local bird populations and migration patterns. For a student whose family worried about the impact of proposed budget legislation, design work during the Capitol Protest project became an opportunity to share her family's story. Following personal interests deepened students' sense of ownership over their learning, led to increased care about the quality of their final designs, and promoted the overall success of the learning environment.

Not surprisingly, many students referenced choice and ownership when comparing PPS to more typical school– based learning experiences. As one student remarked about the Capitol Protest project:

> *"I think for years and years students have been asking their teachers, 'How am I going to use this in the real world?' Teachers either brush it off or try to come up with excuses. I feel like this is actually very applicable. I think the moment you take something and make it matter in a person's life they'll be much quicker to jump on it and participate. So I think social relevance is extremely important and why this class is so appealing and the flexibility of it. Just because we're doing this project on the protests doesn't mean that we would have to. We could be doing anything. Just the knowledge that you have that power to choose what you are studying and how you're studying makes it a much more hands– on experience. I think that's why it's that much more enjoyable. I think that's why it's important."*

This is not to suggest that students always agreed on topics and designs, or that the decision making process was conflict free. On the contrary, it was not uncommon for the whole class, or smaller groups of students, to confront conflict when negotiating rights and responsibilities, choosing topics of study, and making design decisions. One student mentioned this as one of the most challenging parts of the class:

*"I think the hardest thing [about the design process] is trying to get a mutual consensus with a lot people. You know one person wants to do this, the rest want to do this. It's like, 'What do we do to make it in the middle?"*

As a site of democratic practice the PPS classroom was frequently contested, as students, teachers, and community members deliberated, worked to understand conflicting perspectives, and made shared (albeit not always equal) decisions. Though design experiences characterized by choice and ownership were messy and often time consuming, we believe they are critical because they provide opportunities for students to engage in, rather than simply talk about, democratic practices.

### relationship between Voice and Design

The second theme to emerge across both projects was the centrality of student voice in the design process. Students were eager to share their personal experiences and perceived media design as an opportunity to develop and voice their own stories, opinions, and perspectives. In particular, students were concerned with the representation of youth voices. This expression of voice arose from students' perception that youth were often excluded from local decision–making processes. In response, students approached their design work as a means to educate other youth about local issues and include them in public discourse. In the Capitol Protests project, for example, students noted how their documentation and media products could teach other youth about the protests. As witnesses, many expressed a sense of responsibility to document and share insights and interpretations. One student described the intended influence of the Capitol Protests design by noting:

*"A year from now the kids in this class may have heard about it but they might not have been able to do much. Even in a few years from now, there will be people who are in middle school or elementary school and they heard about it, but they might not know exactly what was going on or have experienced it the way that we are. So to have that way of showing them this is what we saw, this is what was done, this is what the people were saying is a huge thing."*

In design meetings students often referenced youth as the ideal audience for their products. They believed their designs were educationally relevant to their peers and capable of shaping present and future perceptions of contested community issues.

The design process also fostered new interactions with people and places in the community and provided opportunities for student to engage in public discourse and express their opinions. This was evident for one student who worked as a videographer on To Pave or Not to Pave. Reflecting upon his documentation and media design, this student discussed the importance of being able to speak directly with the city administrator and consult community members who used the conservancy:

> *"... pretty much our entire classes' opinions got heard because we discussed it with the city administrator, and then emailing people about it or emailing the city administrator and actually talking to people in the conservancy. Like they'd say their opinion and then you could say yours and I did a couple of times and I know that my voice was heard."*

This student's growing fluency with video further amplified his voice; his video work facilitated contact with additional community members and resulted in media that was included in the final AR story. This student's experience exemplifies a confluence among voice, design, and place. As he developed an interest in local places he used media design to share his ideas, which in turn further immersed him in his community. This student's learning trajectory illustrates the potential of place– based design to develop students' sense of place:

> *"I never really thought about anything before ... I'd go to the skate park and home and work and that was it, and now it's got me more interested in learning about local issues. I don't think I ever would have thought about [the conservancy] though if I hadn't taken this class."*

Students also used design to assert independent thinking and decision–

making. In a unique instance, one student leveraged his media design as an opportunity to demonstrate a personal capacity to form critical opinions in response to others:

> *"I know that personally my dad is against unions so it's been an interesting experience in my house trying to remain objective and dealing with stuff when I am someone who is maybe not necessarily of that same opinion or viewpoint. I think that just further proves that I can have my own opinions. I can handle this. It's not like it's above my head. That's how I felt too about how students are portrayed in the media. There's this image we're getting of being ignorant and people just looking for a good time ... I think one of the things about this program we're working on is that we will give it a much more clear and unfiltered view of how students are actually dealing with the political climate rather than a news person telling you."*

This student's perspective demonstrates how design is closely tied to a sense of self. For him, engaging in design represented an act of agency and a reaction to both his personal experiences within his family, as well as mainstream media portrayals that cast teenagers as incapable of independent thought and easily manipulated by parents or teachers.

### perspective recognition

The third theme related to place– based mobile storytelling and civic participation, perspective recognition, concerned students' ability to examine controversial issues from multiple perspectives. The development of perspective recognition as both a skill and disposition is necessary for participation in a pluralistic society (Barton & Levstik, 2004; Hess, 2009). Significantly, in both projects students had to consider whether their final design would be explicitly persuasive or take on a more nuanced, multi– perspective style. Because they ultimately chose to produce media that represented multiple perspectives and honored a plurality of voices, the students had to actively seek a range of opinions and perspectives as part of their inquiry. This also required them to make decisions about how to effectively and accurately present these multiple perspectives in their final design. Both inquiry and design encouraged students to recognize the nuance and complexity of local civic issues. In some instances this required them to reflect upon their own perspectives, question assumptions, and even change their opinions — all practices that develop a disposition towards perspective recognition.

In To Pave or Not to Pave students reported that design decisions shaped their thinking about whether or not paving the path was beneficial, and why both decisions had value. Throughout the inquiry and design processes, students were encouraged to analyze issues and related tensions, often requiring them to examine data and perspectives initially considered irrelevant. As students imagined how different users might see the controversy, some of their initial opposition towards development softened. As evidenced by the following three quotes, learning content associated with multiple perspectives informed students' opinions as they proceeded through the design process:

> *"I was against the path for a while because I only heard people in our class saying, 'It's ruining this; it's going to do this, it's going to do this.' Then I heard people for the other side and I was like, 'What am I going to do? I don't know any of these opinions or anything.' I read articles and stuff and as I was reading it, they were making really good points."*

> *"I didn't really think about the different aspects like the environmental things such as runoff and washout. I didn't think about those at all. It was more like runners and bikers and I didn't think about accessibility at all either, like people in wheelchairs or strollers. I didn't think about any of that. So it's like there are a good amount of pros that I see ...."*

> *"[I] wasn't thinking about how you know, handicapped people could get around or that maybe bikers should be able to go through even though some people are talking about how it scares away the birds ... so now, I'm kind of like in the middle."*

Aspects of perspective recognition were similarly exhibited by students in the Capitol Protests project. After investigating the proposed budget legislation and related political activity as citizen ethnographers, students chose to design a mobile– based game that was "more objective." This decision required students to consciously distance themselves from personal opinions during various stages of the design process. Intentionally distinguishing among opinions demanded that students not only develop awareness, but also refine a capacity to manage multiple roles and respect divergent viewpoints. Students' experiences as citizen ethnographers reflected this sense of distance and respect for varied opinions:

> *"If you go on your own time, you're going for what you believe in. You're going to stand up. You're going to protest. We didn't go to protest. We went to document. We went to ask questions. We went to see in depth. A lot of people don't do that."*

> *"... it definitely changed my views a little bit and made me want to learn more about what's really going on. I'm still not taking sides on anything though."*

## place– based design education

Both of these projects exemplify a pedagogical and curricular approach that differs significantly from traditional models of social studies and civics education, and students' often uninspired experiences of the discipline. From our perspective as teachers, we believe that People, Place, and Stories presents a unique synthesis of three core values: place– based learning, design– based learning, and democratic participation. While other educators may share similar values, we believe that integrating them into a coherent whole produces a distinct framework that foregrounds design as a means of supporting students' participation in the civic fabric of the classroom, school, and community. We call this framework Place– based Design Education, as presented in Table 1 *(page 146)*.

Place– based Design Education provides unique opportunities for students to reflect on and participate in their local communities. In particular, it foregrounds design as a method for actively critiquing, rebuilding, promoting, and shaping local cultural and ecological traditions. This action, or practice— oriented, approach positions students as presently capable of, and already, engaging in the civic fabric of their community. By situating learning around ill— defined and real— world civic issues and problems, place— based design education promotes democratic participation and citizenship as fluid rather than static endeavors, and emphasizes the importance of learning by doing. Based upon our pedagogical commitments, learning by doing becomes synonymous with participating by designing.

## reference

Apple, M. & Beane, J. (2007). Democratic schools: Lessons in powerful education (2nd ed.). Portsmouth, NH: Heinemann.

Barton, K. & Levtstik, L. (2004). Teaching history for the common good. Mahwah, NJ: Lawrence Erlbaum Associates.

Hess, D. (2009). Controversy in the classroom: The democratic power of discussion. New York, NY: Routledge.

Mathews, J. (2010). Using a studio— based pedagogy to engage students in the design of mobile— based media. English Teaching: Practice and Critique, 9(1), 87— 102.

Mathews, J. & Squire, K. (2009). Augmented Reality gaming and game design as a new literacy practice. In K. Tyner (Ed.), Media literacy: New agendas in communication (209— 232). New York, NY: Routledge.

Smith, G. & Sobel, D. (2010). Place— and community— based education in schools. New York: Routledge.

Squire, K. (2009). Mobile media learning: multiplicities of place. On The Horizon, 17(1), 70— 80.

TABLE 1: FRAMEWORK FOR PLACE— BASED DESIGN EDUCATION

| Core Value | Relevance to PPS | Student Voices |
|---|---|---|
| Place— based Learning: an educational approach that emphasizes the study of local cultural, ecological, economic, and political systems (Smith & Sobel, 2010). | Place allows students to draw from their personal experiences and provides a real— world context for learning. Students' inquiry and design experiences change their perception of their local community and foster a sense of place. Students care about and can see the immediate results of their work. | "I think if you're living in a certain area, you should want to be a part of that and make it the best living space for you, for your family, a safe community and a community that you want it to be ... I felt if you speak up for something you believe in, you're putting in that effort to make your community what you want to live in." |
| Design— based Learning: a constructivist approach that engages students in defining and solving open— ended challenges, through iterative, non— linear cycles of design and inquiry in which they envision, build, and evaluate products, events, and learning experiences. | Design guides students' learning experiences and engages them in creating media that is personally relevant and reflective of their own interests and experiences. Students' design experiences open unique opportunities for interacting with one another and place and deepens their understanding of core content and concepts. | "I think the designing part is where you learn the most. I think we got more out of it than a regular class. But, if we designed it and then played it [ourselves] it wouldn't have been any fun because we were trying to teach other people about an issue that we already knew about." |
| Democratic Participation: an approach that emphasizes students' right to make decisions about their own learning trajectory and fosters student voice through deliberation and shared decision— making (Apple & Beane, 2007). | Democratic participation cultivates student ownership over learning in the classroom and community. Students co— design the course by shaping the questions, issues, people and places we study. | "Like after we finished something and he says what would you like to do next? Would you like to work more on this? Would you like to move on? ... We had the option. If I really wasn't into something very much, I could do a little bit more work on it, but if I wanted to move on, I felt comfortable enough raising my hand and saying, 'you know, we've done a lot with this, I think we should start moving forward more'." |

# re:activism:
# serendipity in the streets

*by Colleen Macklin*
*Thomson Guster*

# re:activism:
# serendipity in the streets

*by Colleen Macklin*
*Parsons The New School for Design in New York City and Director of PETLab*
**and Thomson Guster**
*Kelly Writers House*

What happens when the world of a game gets confused with the world? Re:Activism is a location– based urban game about activism. It takes place where most activism takes place, in the streets. Since 2008, the game has been played in New York, Beijing, Minneapolis/St.Paul, and, most recently, Philadelphia. During that game, players on the streets of Philadelphia recreating historic protests encountered and became part of the Occupy protests; two actions intersecting and creating a new form of situated learning, one where the lessons of the past became critical to the present — in other words, a history teacher's dream. This event also raised a number of important questions about the design of serendipity in games, the nature of the interactions that can be created in public through games, and, most importantly, what's learned when playing these kinds of games. In addition to using the Philadelphia Re:Activism game as a case study, insights from game's design and player interviews reveal pedagogical object lessons about what kinds of things are actually learned through place– based play.

## INTRODUCTION

"It was the moment for me where it turned into not being a game anymore."

> *— A player describing a moment in the game when an interview with a Vietnam Veteran raised questions about the nature of the activities in the game.*

Re:Activism is a game originally designed in 2008 for New York City's "Come Out and Play" Festival by PETLab at Parsons The New School for Design. Since its NYC debut, it has been adopted by different educational and activist groups and adapted for play in Minneapolis, Saint Paul, Beijing, and Philadelphia by groups of middle– schoolers, college students, and adults. In this chapter we'll focus on the Philadelphia edition of the game, in particular those moments that blurred the boundary between the game world and the real one, and the resulting insights from the game's players. Through this example we hope to shed light on the power of serendipity in location– based games and situated learning as well as provide a truly honest assessment and of what is learned in the game.

In addition to situated moments of learning, the game involves actions that are performed, rather than information which is read or otherwise received. Whether players are chalking facts about US war veterans outside of the historic Betsy Ross House (where, on December 26, 1971, members of Vietnam Veterans Against The War protested to call attention to ongoing US war crimes in southeast Asia), encouraging passersby to join them in singing "Age of Aquarius" on Independence Mall (where, on April 21, 1970, the cast of the Broadway musical Hair performed at the first ever Earth Day), or creating a historical marker to commemorate the Dewey's Lunch Counter demonstrations (where, in April and May of 1965, the management's denials of service to LGBT patrons triggered a concerted protest that caused the management to recant its discriminatory policies), players re– enact and commemorate historical events, intervening provocatively in public space, playing teachers in the streets, educating passersby about the ways the past lives on in the present moment. The connection between embodied experience and learning in dynamic public spaces ultimately leads to outcomes that exceed the designer's control, where experience and serendipity (defined below) combine to make the past more immediate, allowing history's lessons to be understood in light of the present moment. In this chapter we'll trace the contours of these unpredictable outcomes and describe the challenges and possibilities in designing learning that is active, serendipitous, and unpredictable.

## WHAT IS RE:ACTIVISM?

Re:Activism is a location– based urban game that maps the history of activism onto the public spaces where they occurred, as players reenact and re– create the actions that once took place there. Re:Activism focuses on the actual practices of activism to reveal not just the what, where, and when of water-shed protest events, but also the how, and it does this by enabling the players to interact with the public through playful re– enactments and other interventions. Teams compete by racing between sites of historic protest and activism, earning points by completing challenges at each site that not only "reactivate" the issues represented by those historic struggles, but also serve as occasions to interpret and appreciate anew those who actually lived them. Teams can visit as many sites as they wish, in any order they choose, completing only those challenges that they want to. Though Re:Activism explores history, its primary content is very much of the moment, emphasizing as it does the intrinsic connection between navigating the city and planning and docu-menting public actions.

The gameplay is simple. As the game begins, teams of 3– 5 players are given a backpack containing a map of sites, lettered A through R, of historic protest and activism across the city, with the point value of each site clearly indicated; a packet of sealed envelopes, numbered 1 through 18, one for each site; and other supplies necessary to complete game challenges, like posterboard, markers, sidewalk chalk, and tape, for example. We make sure that at least one player on each team is equipped with a cellphone capable of taking and sending pictures and video — an increasingly common occurrence! Over the course of about 3 hours, these teams race each other from site to site, completing chal-lenges at each location in order to score points. Each site challenge recalls the history of that site, requiring the players to reenact, commemorate, or symbol-ically continue the struggles of activists past — but, before they can attempt these challenges, they must "unlock" each site by answering a "key question," a question that can only be answered by physically investigating the site. For example, at site N, the site of the MOVE House Bombing: "In front of

6221 Osage is a sign denoting permit parking for whom?" Players send their answers to "Protest HQ," a group of game referees with the list of answers to the key questions, who unlock that site when sent the correct answer ("Philadelphia Police Civic Affairs," in this case), texting the players to reveal which of the numbered envelopes in their pack corresponds with that site.

Opening the appropriate sealed site envelope reveals information about that location's history and the site's challenges. For example:

### Site N:
### MOVE House Bombing

**When:** May 13th, 1985

**Where:** MOVE House, 6221 Osage Avenue

What: In 1985, the compound of the black power organization MOVE, located on a residential block in West Philadelphia, became the site of an infamous confrontation, the specifics of which are hotly disputed to this day. What is clear is that when the police, who had come to the compound to serve search warrants, were denied entry, a chain of events was set in motion that culminated in the police bombing the house, killing everyone in the building except for one woman and child. The fire from the bomb spread to 65 nearby houses, effectively destroying the neighborhood. Though courts eventually ruled that the police had used excessive force, no one was found guilty of any criminal wrongdoing, and no jail time was served. The victims of these attacks and their supporters contest the validity of the "official" version of events to this day, demanding reparations from the city for its undeniably brutal and heavy– handed attacks. To this day the Philadelphia Police Department remains the only such department in the country to have bombed its own citizens.

After the site's historical significance is detailed, the site's challenges are presented:

**For 200 points:** Ask a passerby if they either remember the MOVE House bombing or have heard about it. Ask them to explain the effects the bombing may have had upon their neighborhood. Document with video.

**For 400 points:** Currently the MOVE House does not have a historic marker erected by the Pennsylvania Historical and Museum Commission. Using the materials provided, create a sign marking the house and the eleven lives lost in the bombing. Document with a photo.

**For 600 points:** The problem of excessive police force has always been controversial, especially when the lives of the officers are thought to be in danger. Converse within your group about what the police and the members of MOVE could have done to ensure the safety of both the citizens and the police force. Document with video.

**For 800 points:** Draw a series of chalk arrows connecting the site of the MOVE bombing to Philadelphia City & County (the closest police station) located at 5510 Pine Street, in order to emphasize the historic and ongoing tensions that characterize the relationship between the police and the communities they are meant to serve and protect. Document with video.

Teams decide which challenges they would like to complete, and in what order. Throughout the game, Protest HQ keeps track of each team's progress, keeps all teams informed of their standing on the scoreboard, and keeps track of the time remaining in the game. Players send cell phone photos and videos as proof of their accomplished challenges to Protest HQ in order to score points.

This documentation primarily serves as proof of accomplishment for score–keeping purposes, but it also serves as an archive of in– game experiences, providing touchstones for post– game discussions that synthesize the different experiences of each player and each team, and provides valuable feedback to the game designers that can be used to improve and modify future iterations of Re:Activism.

(Indeed, this archive may prove to be a great educational resource for a variety of learning settings — a way for students to feel connected to their city's history, a way for students to become teachers of that history for others, a way for teachers to excite and motivate future students toward their own learning by showing them the media archives of previous Re:Activism participants, and so on. For historical incidents that have already been substantially filmed and photographed (like, for instance, the MOVE Bombing, which took place live on network news), comparisons of extant media documentation to their own Re:Activism archive may help players to feel their connection to history even more deeply.)

Teams report to Protest HQ as they complete challenges. Protest HQ tallies up the points earned and announces the team's total score to all players, functioning as the referee and the scoreboard in order to keep the teams apprised of each other's progress and stoke the fires of competition. After each completed challenge, the team must decide whether to complete additional challenges at that site or whether to move on to another.

The sites featured in Re:Activism Philadelphia were mostly clustered in a few neighborhoods — Old City, the eastern portion of the city famous for its colonial– era historic sites, and Center City, which holds many cultural and civic institutions. Because of the density of Re:Activism sites in those areas, teams may have foregone visiting the outlying — but no less historically significant! — game locations, like the site of the MOVE bombing in the far west of the city, if the game designer had not incentivized such visits by increasing the point values of to those sites' challenges or by adding bonus challenges.

Ultimately, though, each group devises a strategy early on: do we try to cover as much ground as possible, or do we stay in one location and try to complete all of the challenges there? Providing different strategies and locations gives each team a very different experience.

## WHY ACTIVISM?

Activism in the game is a marker for an embodied moment in history — exclamation points marking longer struggles and critical issues of the day. In this way, Re:Activism is a game about history and the role of civil disobedience, strike, riot, and protest in marking history. Reactivating these performances creates an experiential link to historic actors and provides insight into the issues of that time. It also introduces players to the public and spatial nature of activism, perhaps changing the way a street corner someone passes every day is viewed after actually re– enacting an event that happened there in the past. Activism is history performed live and on location, deliberate interventions into politics from outside of the traditional political structure — activism is literally about changing the game society is playing.

## HOW DOES IT WORK?

The technology behind Re:Activism is pedestrian — literally available to everyday people walking down the street: a mobile phone with a camera and the ability to text message. Non– digital tools such as poster– board, sidewalk chalk, markers, and pamphlets act as props and commemorative tools to mark sites and leave behind messages. There's no special software or GPS needed to play the game. This is intentional for three primary reasons. First, it makes the game easy to run by non– technologists. To run the game all that's needed is the ability to send and receive text messages. It's possible to use only simple cell phones as the primary technology, enabling play in parts of the world where computers and internet connections are scarce, but text– enabled phones in abundance (this includes many countries in Latin America, Africa and Asia). That said, when supporting a large number of players it makes sense to use a computer and a freely available text messaging gateway. For the Re:Activism game in Philadelphia we used the freely available online application Google Voice to manage sending texts out to multiple teams. It also made writing the texts easier, since we could type them out in advance and copy and paste them into the application.

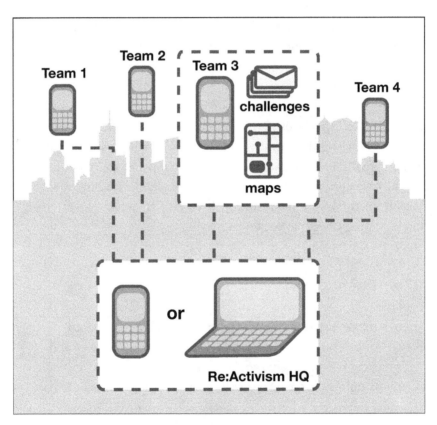

DIAGRAM OF RE:ACTIVISM TECHNOLOGY SETUP

The second reason the game avoids the usage of GPS is to allow the game to be run in dense urban environments where GPS signals often get cut off or are inaccurate. We've found that GPS is almost impossible to use in many areas in New York City due to the concrete canyons formed by tall buildings which allow very little access to skyborne satellite transmissions. Knowing this, we decided to keep the technology (and chances for technical breakdowns) limited, and focused instead on using a combination of texts, paper maps and sealed, coded envelopes to deliver the game's content. Distributing critical game information in different formats creates a reason for players on a team to fulfill different roles — one might be the keeper of envelopes, one the navigator, and the player in possession of the phone, the communicator and documentarian (through texts, video and photos). In early prototypes we used the phone's mapping software, but quickly realized that nothing is more disconnecting than a game that requires players to stare into screens and not look at the environment around them.

Finally, this low tech approach replicates the way that cell phones have actually been used to drive activism and protests the world over. From the famous text– sparked protests in Manila in 2001 that peacefully overthrew Philippine President Joseph Estrada (Rafael, 2003, Shirky 2011) to other examples around the world, a simple text message is a powerful tool for organizing groups and communicating during protests. Phones aren't just used to organize groups, they're also an important reporting tool, providing images, videos and a record of the event from the inside. Twitter and other SMS– based broadcasting services often report mass actions earlier than the news media does, and becomes a source of ongoing news directly from participants as well as reporters, demonstrated most recently by Twitter's reported top hashtag of 2011: #egypt, used to mark posts about the protests there that ignited the Arab Spring. Using text and the phone's basic capabilities is a lesson in how to report and communicate important events.

Not all location– based games need to utilize GPS or specialized applications. It's possible to create a meaningful situated experience through minimal, or even no, technology. However, the technology used can also provide a learning opportunity. What are the digital literacies embedded in the project, and how do the mechanics of the game reinforce them? This is often overlooked for the game's content, and the assumption often is that that the learning involved in location based games is mainly learning about the game's content. However, in our repeated experience with Re:Activism, this couldn't be further from what actually happens.

## WHERE'S THE LEARNING IN LOCATION- BASED LEARNING?

Since 2008, PETLab has run the game in different locations and with different players. We've conducted formal assessments (through survey and pre/post game interviews) and have hosted conversations with political scientists and historians. Each time, one thing became clear: instead of learning and retaining information about the history of activism and of the city, players learned a host of other things one would not link to conventional understandings of the content. They learned how to approach strangers and ask them questions, they learned how to use mobile phones to report and document, and they learned how it felt to perform actions they would usually not perform, in public. If the game is meant to teach the history of activism, it does not do it very well. Instead, it seems to teach some of the basic tools necessary for civic action while using history as a set of "prompts" or learning occasions.

Broader themes emerge and a general understanding of activism may be solidified during post– game debriefings, but, at heart, the game's core mechanic — a race — is in opposition to the patience and careful reading necessary for players to gain much more than a surface familiarity of the historic events that underpin Re:Activism's challenges. "The mechanic is the message" is the title of a set of non– digital games by game designer Brenda Brathwaite. In the games, Brathwaite attempts to convey the meaning and emotional impact of historic social tragedies like the Trail of Tears or the Holocaust through simple game mechanics — in other words, what players do. "The mechanic is the message" is also the idea that what players do in a game is what they retain from the experience. Many remember laying on the ground on Wall Street when enacting the 1987 Act Up "die– in," but they might not remember the details, such as the date, the name Act Up, or even what it was about. However, we believe that even if this information is lost, there's something else gained. We use the terms "collateral learning," "stealth learning," and "serendipitous learning" to explore what kind of non–traditional learning occurs in the game.

## STEALTH LEARNING

In the same way that an arcade style game about math might teach more eye–hand coordination than actual math concepts, Re:Activism teaches actual practices and tools used by activists and the affects involved in public action. We call this stealth learning. This kind of stealth learning is found in the game's mechanics — the actions players perform. Chalking messages in the sidewalk, forming a human chain, creating slogans and placards, chanting, interviewing members of the public, texting, tweeting, recording and photographing. Using mobile phones and social media to delve into the histories of activism allows players to access history through performance by allowing them to inhabit the role of an activist. At the same time, it permits them to inhabit the role of the "citizen journalist," the documentarian who can use such a simple thing as a phone to spread and capture news in text, photograph, and video, reporting live as events unfold.

The performative aspect of the game is a way for players to learn that may not be recognized as such — hence, "stealth learning." But, as they repeatedly engage in these behaviors during the game, they become savvy with a variety of skills: navigating through their city, organizing and collaborating within the group and with strangers, and intervening in public space using the theatrical and technological skills mentioned above. Whether there's a long–term retention of, or an increased disposition toward, any of these practices has not been assessed. However, just as videogames develop eye–hand coordination and new forms of digital literacy, locative games too have the potential to cultivate new skills with technology and spatial awareness. More interestingly, the game may foster a new dispositions and attitude toward the historical and activist content of the game, or change the player's relationship to their city, prompting a re–evaluation of the player's place in it. We borrow from John Dewey's notion of "collateral learning" to explore how Re:Activism operates in this way.

## COLLATERAL LEARNING

In Experience and Education, John Dewey describes a form of learning he terms "collateral learning":

> *"Perhaps the greatest of all pedagogical fallacies is the notion that a person learns only the particular thing he is studying at the time. Collateral learning in the way of formation of enduring attitudes, of likes and dislikes, may be and often is much more important than the spelling lesson or lesson in geography or history that is learned. For these attitudes are fundamentally what count in the future. The most important attitude that can be formed is that of the desire to go on learning." (49)*

One of the first things we learned when we talked to and surveyed players after the game was that the race created an urgency that made retaining any of the facts about the site and the history impossible. After a session of the game in NYC for middle– schools, when asked what they learned about a particular site, some responded: "I learned where it happened. I always thought it happened further downtown." Others (more than we hoped) responded like this: "I learned something but forgot it."

We learned through these post– game interviews that what's retained is not a clear understanding of the particulars of an event — the where, when and who — but instead what players did during the game and the attitudinal strate-gies they devised to complete the challenges. One player, in responding to the question: "What did you learn about this event" responded "A lot. All about approaching people, being positive, trying to get their answer, interact." In many ways, players learned the attitude of activism over its history. Learning in the game was certainly collateral, and we would argue that games themselves are not the best tool for teaching facts and data. Instead, games convey attitudes, strategies, and emotional messages through the fact that they are experienced as dynamic and shifting systems that respond to our actions, rather than static containers to refer to. This also leads to emergent and unpredictable outcomes, ones that can't be controlled, but can be leveraged through design. Leaving room for serendipity and coincidence is one design strategy.

## SERENDIPITOUS LEARNING

We believe that games of this type demonstrate truly meaningful – but not conventional or "testable" — learning outcomes. Because they take the form of a compact event, players are able to suspend the concerns of everyday life and try on new attitudes and roles in public. At the same time, they're given license to interact with the real world in ways that they might not have without the game. This enables serendipitous learning, the kind that is remembered as an authentic experience, not just as information about a given topic. In other words, the "wildness" of the game, due to the fact that it is played in a dynamic public, leads to stronger impressions than more contained experiences. In Philadelphia, players of the game playing activists encountered with actual activists, joining in and learning from the Occupy Philadelphia movement during its crescendo. In fact, an entire team defected from the game to participate in this, as they called it "once in a lifetime opportunity." Would they have joined the march if they weren't out already, playing together as a group? When asked that question, the answer was probably not. By simply making oneself available to the possibilities of the moment, players found themselves doing something new and unexpected that they will always remember.

In addition to this event, other players met and interacted with people who had a real connection to the site or the issue that that site represented in the game. Many of the activities in the game involve interviewing members of the public. One player, when recreating a protest against the Vietnam war coincidentally found a Vietnam veteran to talk to. He had been blinded in the war, and before the interview began, asked the interviewer if she was going to spit on him. When she emphatically said no, he went on to describe his experience returning from the war, when people spat on him and called him names like "baby– killer". He argued that the anti– war protests of the time were often organized by those uninformed about the actual experiences of the soldiers who went to fight. The reception he received on his return, injured and blind from a war that was unimaginably brutal, was an additional assault. His perspective on protests that the player would have agreed were "right" led her to question her actions in the game.

She says:

> "It was the moment for me where it turned into not being a game anymore."

> "I started questing more and more after doing certain tasks whether I knew enough about what the tasks related to for me to be able to judge for myself how much I agreed with what I was doing. It definitely brought a new level of reality to the experience."

That level of reality was only made possible when the game world and the real world intersected. It was the moment for that player and other players who ended up leaving the game and joining the actual Occupy march, that the relationship between history and the present came into focus. It's when re– enactment became practice for something with higher stakes. And it created experiences that will also likely be remembered far more than the facts written down on the Challenge Cards for the game. Serendipity can be cultivated in a game. The most direct way is to include interaction with non– players, eliciting responses and diverse perspectives on the game content. This is how Re:Activism generates serendipity. In past games, players have encountered members of the public that even had first hand experiences with the events, providing an even greater level of depth into the content. The first time the game was played, at the 2008 Come Out and Play Festival in New York, one of the teams met women who were present during the Stonewall Inn Riots, and proceeded to enjoy a beer with them at the site (CITE Macklin, 2010). These moments transcend the game, and became for those players a powerful memory, part of their own personal history.

## WHAT WE'VE LEARNED

**design for chance:** Part of the reason that players met Occupy protesters was because the game designer added Occupied City Hall to the list of game sites and included challenge that asked the players to speak to, document the activities of, and otherwise interact with the Occupiers and the police who had been deployed to monitor them. We wanted the game to demonstrate that, in the words of William Faulkner, "The past is never dead. It isn't even past." Activist struggles persist to this day, and the battles once fought are still being fought. By choosing to include sites that represented events from the earliest days of colonial America all the way to the day of the game itself, we hoped to help the players find that blurry boundary between the game and the real world, to discover that the dramatic public interventions the game required of them were akin to the dramatic public interventions that current events demanded of the protesters. Designing for chance, in the case of Re:Activism, means attending to the contemporary context in which the game will be played just as much as the historical backdrop that provides the bulk of the game's challenges.

**design for openness:** Each team set about playing a different way. One team focused on acquiring as many points as possible by visiting as many sites as possible, and so they stuck mostly to the more densely clustered sites of the Old City area, making strategic trips to outlying sites in order to capitalize on the bonus points there. Another team picked out locations they wanted to personally learn about, weaving a more circuitous route through the city in an effort to satisfy their particular curiosities. And another abandoned the game entirely to join the Occupy Philadelphia marchers, where they learned about activism beyond the game's parameters. By striking a balance between emphasizing the rules of the game on the one hand and the spirit of the game on the other, game designers can help create permeable or fuzzy boundaries to the game world that allow players the greatest potential for rich experiences. And there's a deeper lesson here, too: activism, being a necessarily social enterprise, is not simply a way of demonstrating knowledge or taking action, but a way of acquiring knowledge, starting discussions, and it rewards those who take ownership over their own learning experiences.

**give players tools to help guide chance encounters with the public:**
Players were given Re:Activism shirts to wear, shirts that fulfilled a few purposes. First, they provided a "point of entry" for passersby to approach the players as they played the game, performing reenactments, posting signs, or otherwise presenting an unusual spectacle in public. This "point of entry" function also helped players to easily identify themselves to passersby for those tasks which required them to interact more directly or conversationally with the public. Second, the shirts identified the players as members of a group, and, in particular, as members of a non– activist group, giving them some modicum of "cover" or license to behave unusually in public — to, in some cases, post signs or chalk on historic sites or accost passersby — without attracting any negative attention from security guards or police. In these t– shirts, they were friendly, recognizable, good– natured. Third, these branded t– shirts perhaps made passersby curious about what "Re:Activism" was, or at least allowed for that possibility. Lastly, and most importantly, these matching shirts, by setting them apart from the rest of the public, gave the players the push that they might have needed to actually go out and behave unusually.

The packets of historical information in every team's backpack also served as a useful tool for guiding players through their encounters with the public. Providing them with summaries and talking points about a variety of complicated historical events, these packets give players the authority they need to cold start conversations that might not otherwise occur informally, in public. And these history lessons are brief, bite– sized, perfect for consuming on— the— go as teams raced each other through the game, a nourishing supply of food for thought to fuel conversations that roam across topics even as the players roam the city.

**plan a route and playtest it!** The routes were play– tested by small groups to make sure that no historical sites were closed or under renovation; that there was enough room at sites to allow for performance– based challenges; that there were enough nearby poles, walls, benches, etc. to which posters and signs could be affixed; that relative indicators of orientation in the game directions were replaced with absolute ones (for example, the "east side of the building," not "the left side."); that the addresses and street intersections for all the sites were accurate — or, if there was no obvious address or intersection (as was the case for the site of the Black Bottom neighborhood protest) that the site was indicated very specifically by landmarks; and that the key questions were all answerable and that the answers were all recorded correctly.

Certain events that made it on to Re:Activism's map took place at sites that no longer exist (the entire Black Bottom neighborhood was razed and built over) or have since relocated (the Institute of Contemporary Art, from which the players departed in the morning, was once located in a different part of the city, and it is this previous site that the map & challenges refer to). Making these facts apparent to players was important, but equally important was plan-ning for their possible confusion by knowing in advance which sites may cause confusion.

The packets for two sites, one commemorating the Philadelphia General Strike of 1910, the other the citywide protest against the transit workers' strike of 1944, were unintentionally conflated while we wrote up their respective challenges. The wording of one challenge seemed to demand that the players travel to another location quite far from the actual site in order to earn points. At this point, Protest HQ more than proved their worth when they received a call from a stymied team, clarifying the matter before the players got too frustrated. This incident shows both the necessity of thoroughly play–testing the game route and materials and of being flexible in the event of the inevitable surprises.

**Mix younger players with adults:** Re:Activism's been played by middle school social studies classes and younger children after school. The game works best with players age 12– up, due to the content complexity and maturity to understand and play by the rules of the game as well as the social rules inherent in public performance and play. Mixing adults with younger players provides an increased level of supervision and safety when interviewing strangers on the street and performing the challenges. It also has the benefit of providing cross– generational dialogue on the issues explored in the game and may help to prevent the activities of the players from being dismissed or ignored as simple shenanigans.

## WHERE TO GO FROM HERE?

While the stealth, collateral and serendipitous learning outcomes from Re:Activism are useful and in some ways more vibrant, the shortcomings of the game in relaying detailed factual content does still frustrate some players who were looking for a more deeply informational experience. This could be partially resolved by creating companion pieces around the game, a website to share information and content, and even working with teachers to create curricula that augment the experience or embed the game in curricula that already exist. Partnerships with educational and arts organizations, such as the partnership with the Kelly Writers House at the University of Pennsylvania and the Institute for Contemporary Art in Philadelphia, lead to opportunities for the game to be one of a series of events including teachers, students, activists, and other members of the community. With additional planning, local political grassroots organizations, perhaps even city departments, could be brought in as partners, too.

An opportunity also exists to create a curriculum out of the activity of preparing the game in one's own location. Research, talking to community— based organizations, mapping and learning the iterative methodologies of game design could all be part of a class, after– school program, or learning module. In other games designed by students at PETLab, we've found that making a game about an issue leads to a deep understanding of the systems underneath the issues. Games, as a cultural medium of systems, lead to a greater understanding and facility with systems thinking. And when making them, the importance of systems becomes all the more apparent.

## references

Dewey, John. 1938. Experience and Education. New York: Collier Books

Friedman, Uri. 2011."The Egyptian revolution dominated Twitter this year". Foreign Policy website, 12/05/2011. URL: http://blog.foreignpolicy.com/posts/2011/12/05/the_egyptian_revolution_dominated_twitter_this_year

Rafael, Vincente 2003. "The Cell Phone and the Crowd: Messianic Politics in Recent Philippine History," Public Culture, v.15, no.3, Fall, 399– 425.

Macklin, C. 2010. "Reacting to Re:Activism: A Case Study in the Ethics of Design,"
in Ethics and Game Design: Teaching Values through Play, D. Gibson and
K. Schrier.eds IGI Global

# history in our hands: mobile media in museum adventures

## by Seann Dikkers

# history in our hands:
# mobile media in
# museum adventures

*University of Wisconsin — Madison*

A special thanks to Dan Spock, Wendy Jones, Jennifer Sly and the whole crew at the Minnesota Historical Society for their vision and investment in mobile media learning. Ultimately this is their project and I'm acting here as a journalist telling their story.

## A CAVE– IN AT THE MINE

Anton Antilla worked as a miner on the Iron Range for years. He regretted coming to America from his homeland of Finland, but made a living feeding ore to the industrial giants on the Great Lakes. Through the mobile media game Our Minnesota you are able to scan and talk to Anton. Because you are in a museum, Anton can invite you to enter the 'real' mine directly in front of you, try out the different tools, and consider which job you may want to have. While in the mine, however, a cave– in interrupts your tour and you are called on to save the miners. In the end, the miners are saved, you are the hero, and hopefully you've learned a bit more about history by playing the mobile game Our Minnesota.

The Minnesota Historical Society (MHS) is already known for their exhibits. In past projects like the Tornado Room, D– Day Bomber, and If These Walls Could Talk, MHS sought to give visitors simulations of real events using sound, sight, seat 'kickers', and great storytelling. Anton is only a part of

MHS's newest effort to amplify designed spaces — this time using mobile media learning to place you in the middle of the story. By using mobile devices, Our Minnesota layers narrative and media on top of already compelling spaces for an even better learning experience. Characters, like Anton, encourage immersive role– playing by inviting you to complete quests, gather items, and solve puzzles, while physically walking around 14,000 square feet of exhibit space.

My own involvement in this project came after MHS invited me to join them as a gaming media consultant. They simply asked, "How would you like to build an augmented reality game in a space that is designed for it from the ground up?" Over lunch we talked about what it might mean to design a space with a game in mind and/or design a game with a space in mind. History is full of great stories fueling many of the top games on the market, so the match was made. We began designing a quest driven role– playing game using mobile media devices within a curated space.

Anton is one of many characters in Our Minnesota. Each major region of the state is roughly mapped into the exhibit space. For each area, or 'hub', (including the prairie, metro, forests, iron range, and lakes), we developed a primary character to give quests and secondary characters that populate scenes and stories — all of which are historical personalities. In addition, each hub includes a full scale set piece (sod house, street car, trading post, mine, and port) that can be integrated into the mobile media game design. Each hub is designed for 10– 20 minutes of play time — just enough to introduce the characters, topics, and challenges relevant to that space and time. Finally, while the space can remain constant, the digital aspects can be rewritten as often as MHS wants to — saving resources and time.

This project is unique in that it combines decades of expertise in building interactive spaces with current game design models for engagement, motivation, critical thinking and collaboration. MHS is tackling a laundry list of design questions, using rapid– prototyping, play– testing, and iterative design to inform the design process. Though MHS is incorporating a group of designers, the lessons gathered can inform any mobile project.

## DESIGN GOALS

Our first goal was to think about the use of space in relation to a mobile game. Where a traditional museum has a 'snake trail' that has a distinct start and finish, we could allow the game itself to direct traffic along invisible paths. This eliminated a need to guide visitors with glass case 'walls'. With the different hubs, it was a benefit to have multiple entrances to allow the group to naturally split up and gravitate toward characters that appealed to the learner. The exhibit is also loosely modeled after the State of Minnesota to include northern forests, a metro middle, southern plains and seven other vignettes. These are arranged in an open floor plan so visitors can 'tour' the state, learn about the geography of the state, and even tell their own stories using the floor plan itself.

We wanted flexibility to design new stories and characters over time. Each 'zone' has a primary quest and side quests that allow you to be part of the story, challenge your thinking, and engage you with periods and places. The hub model allowed redesign to happen in much smaller chunks. By tracking student choices, we have already been able to weed out characters and stories that don't attract the target audience and try new iterations. The digital flexibility not only provides for redesign potential, but adds the potential of special events, student game design, and expandability to other exhibits and locations at a low cost.

In terms of the player, MHS wanted the field trip to provide resources to student visitors for use back in their classrooms. Field trips have a long–standing tradition of providing rich experiences. MHS reported that the rising importance of standards has caused some teachers to pull back on field trips because they aren't necessarily linked to standards. By sharing the vast resources of the historical archives, MHS felt they had something more to offer these teachers. Using the mobile devices, the experience of the field trip could be a launching point for classroom projects, discussions and lessons. At best, this meant the design needed to:

- Be engaging to the player
- Introduce characters virtually that would be valuable assets for the teacher
- Introduce historical frictions that would contribute to meaningful discussions
- Provide digital media assets to the players that they could access at school
- Be informed by the history standards
- Involve the teacher with ancillary materials and ideas for the classroom

Our Minnesota is therefore designed as a game that introduces concepts and engages the learner in topics by having them gather resources. For instance, while playing the stories, you can also use the mobile device's capacity for collecting digital images, audio clips, and notes for later use in the classroom. For example, when you find a raccoon trail in the forest, you can photograph it; share it with friends, family, or teachers; use it in a presentation; or keep it as part of an online profile shared with other history buffs. Not every player will meet Anton, or find the trail, but they will all have stories to tell and digital media to help tell it. In fact, the idea that not all students would have identical experiences was both a game design choice and a classroom design choice. If each player has different knowledge to bring to the classroom, it opens possibilities for the teacher and naturally discourages some sort of tested assessment.

Our design constraints were two–fold. First, we wanted players to experience the whole facility so MHS wanted to stay within a 20 minute time constraint. Museums typically design for visitors to be in each exhibit roughly 20 minutes. Early play–testing, however, demonstrated that players were actively trying to finish the mobile game and do multiple hubs in one visit. Post–interviews had students wanting to come back and "spend the whole day playing all the characters." The '20 minute rule' was being broken easily, yet hours in one space meant less time in others.

Second, MHS wanted to use all real characters and stories without losing the value of the space too. Specifically, they put a lot of pride in award winning museum inter– actives, multi– modal simulations, and carefully designed visuals. If the game was "too good" it would pull players into intense focus on the small screen at the expense of the space itself. The game needed to be engaging, but also direct the player to look around, explore, try things, and notice details around them. More on our findings on this later.

## DESCRIPTION OF IMPLEMENTATION

Creating Our Minnesota, or the design process, was the first like it in my experience. I believe that we were able to bring together the best in exhibit design and game design in a way that warrants a brief overview. Your own mobile game, whether in a curated space or not, may be one that involves a group of experts and designers. If so, this process may be a great fit.

### 1) focus groups
MHS started, before any game design, to talk to their audience. With a small state grant they had focus groups across the state asking, "What should field trips look like for today's learner?" Teachers and students identified a digital gap between home and school spaces, the need for 21st century skill development, and the importance of leveraging digital tools to bridge home, school, and museum experiences together. Students needed opportunities for critical thinking, collaboration, and experience– based learning. MHS, through this process, had clear goals and a grasp of what their true assets were to their audience.

### 2) bring together the right people
The Our Minnesota design team included historians, programmers, visual media experts, carpenters, exhibit planners, teachers, researchers, and administrative oversight. MHS's production team has hundreds of combined years of design expertise and their production pipeline is refined. After seeing emergent research on the power of gaming for learning, MHS sought out researchers and designers

that were making mobile games. My own experience was being approached after presenting at a conference and being invited to lunch the next day. MHS intentionally gathered people that could speak to game design, production, and play– testing; then invited them to be part of the planning team. For the entire project MHS has teamed with Engage designers at the University of Wisconsin– Madison and Gaming Matter, LLC..

### 3) show don't tell

MHS showed off their facilities and resources, and we introduced MHS to the research behind gaming media for engagement, motivation, and learning potential— along with some games. Each team visited the other's facilities, got to know each other, and built relationships that served the project later. We also took a field trip to MagiQuest (a mobile game situated in a curated space — for profit) and MHS held 'game nights' in their homes to try out specific games that exemplified potent models. Our goal was for people to naturally see for themselves that gaming media was effective, fun, and could be part of the next exhibit.

### 4) iterative design

We wanted a rapid– prototyping model often used in game design to avoid costly, time consuming misdirection. We used an open source game development tool (ARIS) for simple designs of quests, character interactions, and inventories; games that could be quickly built and tested. Creative use of duct tape and cardboard boxes built mock– ups of the exhibit to test the game in. Finally, MHS had a steady flow of willing visitors that loved to try things out — an asset for most museums. So, instead of extended discussions about best design, the project leader, Jennifer Sly, could routinely say, "Lets build it both ways and test it!" saving hours of debate in the planning process. This process allowed us to identify player feedback, problems, and assets, as well as discover forms of mobile interaction that worked for us.

### 5) open houses

Finally, MHS has a long standing relationship with educators and the public.

Let me add that if you haven't visited St. Paul, the MHS building is nearly as prominent as the State Capital as you drive in. When a new exhibit opens at MHS, the press reports it. When they are claiming to reinvent the field trip for the 21st century, having open houses is as much to meet the demand as to create buzz. Open houses were strategic opportunities to bring in community voices. When opening their 1960s exhibit, folks could also get a behind–the–scenes tour on what was coming. Already this has provided important design considerations and feedback.

## PLAYER FEEDBACK

Player feedback was the primary source of data for our design goals and planning for future iterations. We found that lessons learned from having players play versions of the game were useful to resolve differences of opinion and to confirm that mobile media learning was indeed worth the effort.

### enthusiasm

First, we welcomed the enthusiasm from players as they tried the game out. We overheard, interviewed, and gathered from focus groups a collection of responses, mostly there was an excitement about the game as "fun," "interesting," and players reported "I would do this a ton of times." We didn't expect that students would want to replay the game even with the same content. Also, "there was a lot to do" in the game and students wanted to return to MHS outside of school time to play through all of the options in the game. This effectively questioned the "20 minute rule." We still wonder if this is simply the initial enthusiasm of a new interface, or will be sustainable over time.

### characters are 'real'

We anticipated that characters may not carry the narrative because we purposefully limited graphic representations to static .jpg images and posters. Despite this, the conversations proved to be sufficient to engage players. Players referred to the interactions and game with intimate statements like, "you can talk to other characters," "like, live their lifestyle," "they would give me something," "you can get more quests by talking to them," and "we could burn buffalo poop!" Much like high budget video games, verbs were used in the first person and digital objects became 'real' ones in the discourse. In one case the story wasn't satisfactory to a young man because, we didn't give the epilogue to each of the characters. In another case, we had a grumpy character that a player didn't want to return to because they "were mean." Having characters be 'real' wasn't a problem in the play— testing. Also, knowing that the characters were based on actual people added a "Minnesota touch" for students.

## exploring matters

We were concerned about the small screen commanding the attention of players away from the space. On this point we do have mixed results: Some students did appear to be looking at the screen more than the space. Redesigns of the game leaned toward smaller, quicker conversations with characters and quests that had players looking around for solutions. This quickly appeared in the interviews as "You got to walk around and do stuff, who wouldn't want to do that!," "I liked exploring everywhere and finding different things." In addition the space was featured alongside comments on the game seamlessly. For instance, it was common to hear:

> *"You can go and explore and while you're doing it, you can see what people had to do to survive; like the sod house, how small it was and you could look around to see what stuff was there."*

Though the small screen carried the story, players were easily relating the story to the spaces they were playing in and connecting the characters to the hubs.

## differentiation

Another observation of players was that the game was engaging for all types of students. Teachers noticed that some students were quick to pick up the game and were more than willing to help others along. Enthusiasm for the game created a collaboration between students of varying receptiveness to the technology. A para–professional (student aide) pulled me aside at one play–test and exclaimed, "I love how she can connect with this and go at her own pace! Look at her, she is loving this." We have found consistently that student engagement with the mobile game and student performance at school are not linked. Certainly there is room for interesting future study in this area.

## the technology was not an issue

We anticipated that scanning QR codes would be a challenge and require some training on the part of MHS staff. This was true with many adults, but not true with most students. One teacher commented that, "They just get it." When students did have troubles, we notice that a friend or another student would quickly step in to 'help' the other. This helpfulness largely fits with descriptions of participatory cultures and digital communities described in other studies (Ito, et al., 2008 & Jenkins, et al., 2009). Collaboration around using the technology was positive and encouraging to the designers.

In addition, we used protective cases and lanyards for each of the mobile devices to protect them and provide log in information. These were not an issue for students and successfully have protected the devices to date.

## learning

Post–game interviews repeatedly demonstrated that students were gaining experiences (reported in the first–person) around which they could retell the lives, challenges, and particulars of Minnesota historical characters. Our goal was not for players to memorize facts, but to have context for 1) different periods of Minnesota history, 2) characters they could relate to, and 3) spaces that inspired interest in each geographical region of the state. On all three points players reported "going back in time" and that "having that experience is cool." Mobile media learning, for MHS, is about 'living' out an event or 'living' with people and being able to talk about both with classmates, teachers, and curators.

## DISCUSSION

Our work at MHS is not just relevant for museums. The context, player–testing, and designed space have provided an interesting context to test how mobile media learning promotes visitor engagement, as well as to test a number of interface design issues. We learned not just about spatial design, but about the design of the mobile application itself. That said, because this is still such a new process, we would invite others to contact us and share what they have done differently, studies that verify or challenge our observations, and how to continue refining the design process with ideas.

Final thoughts for us include:

- The process that worked for us, (above), could easily be adopted by other designers that want to develop mobile games collaboratively.
- During design, clearly establish goals regarding 'screen– time' vs. focus on the space itself. Large amounts of text and the desire to include 'facts' can and will take more screen time and conversely less observation of the space; and vice versa.
- Collaboration appears to be a natural byproduct of mobile media learning. Other reserach indicates this; we also observed that many student— student and student— adult interactions happen whether or not we provide 'quests' that encourage them. Collaboration is part of using the technology, discovering/exploring, and sharing alternative story arcs.
- Like reading a text, good writing facilitates the imagination to fill in the images and production quality. With story based adventures, attention is well placed in polishing the prose. Imagination and role– play made our game 'pop' in the player's mind despite fairly basic interface design.
- Designing for critical thinking requires trusting that the player will use the space. Challenges in mobile media may be too easy (scavenger hunt) or too hard (complex puzzles). These two extremes need careful balance. The fear that not every player will finish a task was assuaged a bit when we saw students 1) helping each other regularly, and 2) looking around for

clues. By placing clues in the space (not in the game), we could help scale the challenges without making them too obvious.

- There is much to learn in terms of having players interact with space. In the coming months we are integrating Arduino technology that can activate physical events when the digital game triggers them. This small connecting trigger will open up many possibilities that will further expand the capacity of mobile media learning.

These of course are only some of the lessons and questions that will guide future iterations. In the meantime, if you are ever in Minnesota, feel free to come and play.

## references

Ito, M., Horst, H., Bittanti, M., boyd, d., Herr– Stephenson, B., & Lange, P. G. (2008). Living and Learning with New Media: Summary of Findings from the Digital Youth Project. . Chicago, IL: The MacArthur Foundation.

Jenkins, H., Purushotma, R., Weigel, M., Clinton, K., & Robinson, A. J. (2009). Confronting the Challenges of Participatory Culture: Media Education for the 21st Century. MIT Press: Cambridge, MA.

# mobile gaming in public libraries

*by Kelly Czarnecki*

# mobile gaming in public libraries

**by Kelly Czarnecki**
*Charlotte Mecklenburg Library*

You've probably heard of the 14 year– old from Utah that used the computer technology books from his public library and the quiet space there to develop Bubble Ball, a strategy and puzzle game app for the iPhone in December 2010. This free game was so well— liked in fact, it briefly surpassed the number of times Angry Birds, one of the most popular apps, had been downloaded (Nelson, 2011). While most youth might not have the patience to write 4,000 lines of code, chances are they own some kind of mobile device and have probably used it for gaming. Though this is an extreme example of use, there are many other ways to learn with mobile media. The question is, *how are public libraries supporting gaming and mobile devices beyond reference books in their collection?*

## QR CODE QUEST

In December of 2011, the Charlotte Mecklenburg Library (CML) in North Carolina paid for staff to attend a webinar presented by the Public Library Association called Cracking QR Codes— What Are They and How Can They Help Your Library? While QR codes are not new, I attended the webinar, knowing what they were and what they looked like but not necessarily knowing ways libraries were using them or could use them. The presenter shared examples of libraries integrating the 2D barcodes into scavenger hunts around the physical library space as well as attaching them to books on the shelves which would link to a promotional video about the book called a book trailer or a read– a– like (another book that is similar in plot or theme). A member of the library's marketing department attended and shared ideas of how we could include QR codes in our marketing pieces, especially with programs that repeat year after year. By referring people to URLs of photos and videos from the previous year's program through a QR code, it's an easy way to bring new interest to the program as well as connect one event to another. The possibilities of using QR codes in the library seemed pretty endless.

Of all the ideas shared, using QR codes to craft a scavenger hunt seemed to make the most sense for the type of public library I work at. A scavenger hunt is a rather typical activity for many public libraries as a type of game used to introduce people to the facility and ways of finding information.

The library I work at within CML, ImaginOn, is a youth– focused facility. This means anyone over 18 is not able to use the computers, programming is done for those 18 and under, and in certain areas of the building, only young people are able to hang out.

Scavenger hunts can be a great group activity since a lot of people can participate at once. There are many groups that come to ImaginOn, whether as a school fieldtrip during the year or as a camp during the summer on a community outing. Scavenger hunts are nothing new to ImaginOn, but integrating them with mobile technologies is. We are located in an urban setting, in downtown Charlotte, North Carolina. Many teens that frequent the library have cell phones they use to listen to music, watch videos, or text their friends. Even if the teens do regularly hang out at the library, they haven't necessarily explored everything as many tend to do the same activity every day such as use the Internet or play video games with their friends. A scavenger hunt can be a fun and quick way to introduce them to some different resources and goings– on at ImaginOn.

## MOBILE GAMING IN PUBLIC LIBRARIES

Gaming in public libraries has a track record for fostering interest in other library resources. Whether it's used as a lure to get people into the building or it happens more organically (i.e. the library offers an activity I'm really interested in, which makes me want to explore and see what else might be available), gaming connects people with each other and the facility itself.

Supporting mobile gaming in a public library setting could be as complex as lending out mobile gaming devices or circulating mobile games to as simple as developing a scavenger hunt that requires the use of a mobile device with a QR code reader. Since everyone won't necessarily own such a device, you may consider supplementing the QR code quest with the clues in hard copy form that don't require a special device to read the code, especially if you think enough people might be left out of the experience because they lack the tools to participate. At the same time, this shouldn't be a reason not to try out using QR codes. Having people work in teams in case some people don't have a device with a QR code reader is also another way to get everyone involved. Depending on your community, many people do have smart phones and are very familiar interacting with information using mobile technologies. Free QR code readers are available for all major smartphones.

The first time teens at ImaginOn participated in the QR code quest, they discovered it themselves. In other words, they spotted the large codes throughout the library, knew how to read them with their phones, and then proceeded to ask staff how it all worked. The library staff gave them a handout with the challenges listed. We explained that hints to the challenges were given in the form of QR codes posted throughout the 2nd floor. Each challenge required an answer. For example, the third challenge reads, "Studio i is ImaginOn's music and movie production space. Find the Studio i QR Code Hint to locate the name of the company that designed the space. Name: _____ Bonus points: What is the general nonfiction number for books about animation? (Hint: there is a bookmark in Studio i with this information)."

The hint was a QR code attached to the outside of the Studio i door. It is a large code with a heading directly above the code that reads "Studio i QR Code Hint".

Depending on your space, you may choose to make your codes more or less obvious. The important thing to keep in mind is that you don't want participants to get frustrated. If locations within your library are well marked with up to date signage, you may consider using those places as the anchor points for your challenges. If you're looking to focus your QR scavenger hunt on something a bit harder to find such as a specific reference book or a section in the library not as well marked, you may want to give some hints along the way such as what it's located next to so that people don't get too frustrated and give up. The goal is probably to have players feel more comfortable and familiar with the space, not run away in frustration! Asking for feedback and suggestions on what they thought of the scavenger hunt after they finished or how it can be improved is a way that can give you food for thought when making your next iteration of the game.

## NOT JUST FOR TEENS

In January 2012, Library Journal, a trade publication for librarians with an emphasis on public libraries, released Patron Profiles, a trending survey of public library patrons in regard to how they use mobile devices in libraries.

Nearly 15% of the 21– 40 year– old respondents (out of 2155 total) reported, "they are using mobile services to help their children with research or to find a book." (Carlucci, 2012). According to Rebecca Miller, Patron Profiles series editor, the 21– 40 age group "are avid users of a wide range of library services and they are early adopters of technology" (Barack, 2012). This age group is also more likely to use mobile technology than other patrons though few respondents to the survey have downloaded library apps because they don't yet exist or are just emerging.

While this report doesn't directly address mobile devices used for gaming in public libraries it does show that there is a relationship between accessing information in the physical space and using mobile technologies to do it. This can be important information for libraries wanting to capitalize on that interest. Some things you might want to consider include:

- **Convenience/just in time information.** By including QR codes in information the library is creating— whether it be on a bookmark, as part of a flyer, or on our programming calendars, we're tapping into the ease and convenience of locating additional information through scanning a code. We're also helping make the connection between the real and the virtual by giving people another resource in which they can find information if they wish.
- **Building community.** Whether it's accessing an app that will lead a library user directly to the catalog via their mobile device or supporting a mobile gaming tournament, we're creating access points for people to connect with information and each other.
- **Marketing of resources.** Finding ways to integrate the use of mobile devices with the library's resources such as through scavenger hunts can be a powerful marketing tool to show that the library is a fun and interactive place to be.
- **Exploration of history outside of the library.** Using QR codes for scavenger hunts in the library is just the beginning. If your library is located in a neighborhood or community rich with interesting historical facts, mobile scavenger hunts can be used outside the building and take the form of geocaching through using GPS units or building an online game by participants using SCVNGR. Finding those access points to resources unique to your library and of interest to the surrounding community can help build information rich settings.

## DESIGNING A QR CODE QUEST

When creating the quest for teens at ImaginOn, we took several things into consideration. We chose challenges that highlighted spaces or resources that were underused. In this sense the quest functioned a bit as a marketing tool to help some areas and resources gain more familiarity. For example, the Reader's Club is an online site maintained by library staff that includes such information as book reviews, new releases and author interviews. Teens don't necessarily know to use Reader's Club as a resource when they're trying to find something to read— whether as assigned for a school project or something recreational.

This challenge linked the Reader's Club site to the physical space of the fiction collection in the library. The challenge reads, "Look for the Reader's Club QR Code somewhere in the fiction section. Write down two of the latest titles that are reviewed in the Teen Corner of the Reader's Club that sound most interesting to you." Their answers can also be entry points to further discussion in getting to know the participants better. It can also help identify if people are just writing down anything to finish or are at least giving some thought about what they're putting down as an answer! If they really didn't find anything on the Reader's Club site that looked interesting, that can be another avenue to find out what it is they read or what kind of games they like to play that can help make connections to other resources the library owns such as downloadable items, magazines, or graphic novels.

Another consideration we took into account was designing a challenge around areas that were well used. For example, the Gaming Corner which has console gaming (Wii, Xbox, PS2 and 3) during the week when public schools adjourn as well as during open hours on weekends, is a popular place in the library. The challenge we integrated with the QR Code Quest was this: "Stop by the Loft's Gaming Corner. Find the Gaming Corner QR Code Hint. What is one of our newest dancing games and what console does it play on?" The challenge was designed to let teens know that we own a game they might not have known about, we do try and purchase up to date games, and their suggestions are taken into consideration for purchase. Consequently, because the Gaming Corner is frequented at the library, teens notice the QR Code Hints and ask us what that's about. This gives us an opportunity to invite them to participate in the scavenger hunt while they're waiting for their turn to game or to try something that might be completely different for them in the library if they pretty much just console game when they're at ImaginOn.

Many sites are available to create QR codes for free. You can choose a search engine and practice with what comes up when typing in 'QR code generator' or look at the features of several mentioned below. Most sites generate a QR code after you put a URL into the online form. A phone number, SMS message or any text can also generate a QR code. Depending on the site you're using, once a code is created, you can take a screen shot and add it to the document you want to incorporate it into. Be sure and check that it works by testing it with your own QR code reader. It should bring you to the information such as a web site that generated the code in the first place.

I regularly use the Kaywa QR code generator (http://qrcode.kaywa.com/) as it gives several options of information to use in order to create a code (URL, text, phone number, or SMS message). It also lets you choose the size of your code. The Google Chrome browser creates a code for any URL that is currently open— you simply right click to generate the code. This works on images as well and can be shared easily with other social networking sites such as Facebook or Twitter. QR Droid (qrdroid.com/generate) is another site that generates a code after you enter your URL. There's also a QR code template available for setting up a scavenger hunt on the Active History site (Tarr, Russel, 1998– 2012).

QR codes are used in all kinds of libraries including school and academic for more than just scavenger hunts. They can be used for anything from linking additional information to library exhibits such as an artist's web site to video or book trailers before material is checked out. The key with QR codes, whether used for a scavenger hunt or other purpose in the library, is to bring additional content to the user. While the challenges for a scavenger hunt may focus on underused or regularly used areas within the library, the key also is to highlight some additional content for the participant. In other words, rather than have the code simply link to the library's web site, choose something more specific that you're aiming to highlight— perhaps content that might be a bit underused. Link the content to something physical in the library so that the connection makes more sense. A QR code that shows the library's electronic holdings placed near the print materials may give the challenge more of a context to show that there are multiple ways that content can be obtained in the library.

## A WORD ON MOBILE GAMING AND PUBLIC LIBRARIES

While this chapter solely focused on the integration of QR codes as a scavenger hunt game, it's fair to say there is a lot of movement in public libraries with gaming and mobile devices. There is also a lot of time and funding spent on other uses of mobile devices with the public library such as apps to interface with the catalog or the loaning out of iPads to toddlers. To go more in depth with any of these examples, is beyond the scope of this chapter, but it is worth mentioning that integrating mobile devices with the public library is very much on everyone's radar and significant strides have been made in combining access and information in this way.

Gaming is nothing new in libraries. Board games have been played for years and video games, while slower to catch on, have become a part of regular services at many libraries. Whether it's allowing games to be checked out or hosting a tournament, public libraries are generally very supportive of this activity. In a 2009 Library Journal blog post of the Games, Gamers, and Gaming column, writer Liz Danforth points out how video game technology is constantly changing. She asks, do mobile games have a place in the future of gaming in libraries? Can your library support games and gaming played on mobile devices as an organized activity, competitively or cooperatively? (Danforth, Liz, 2009).

A few examples of what public libraries are doing support gaming and mobile devices include:

- **Find the Future**, New York Public Library. In honor of the NYPL's centennial celebration, starting in the Spring of 2011 as an overnight adventure, people were invited to download an app to the iPhone or Android that would unlock a clue to an object located in the library. Once the object was found, players would write a short essay inspired by the object which was to be part of collaboratively written book. The game was able to continue to be played online (Find the Future at NYPL: The Game, 2011).
- **Finding History**, YouMedia, Chicago Public Library. Teens participated in a high– tech scavenger hunt using GPS units to locate geocache's throughout the city. The activity was used to engage teens outside of the library around One Book One Chicago events focusing on Daniel Burnham, an architect who helped plan the design for Chicago in the early 1900's. Clues about Burnham and his plan could be located using GPS units (Karp, Josh, 2010).
- **Foursquare**. Topeka and Shawnee County Public Library. Every time players visit the library, they can check in with their mobile device and earn points. Their page also includes a list of activities to do while visiting the library including getting a library card, using a database for getting full– text magazine articles, and playing a video game (https://foursquare.com/v/topeka– shawnee– county– public– library/ 4b06c23cf964a520b6ef22e3).
- **NYC Haunts**. New York Public Library. In 2011, a partnership with Global Kids and the New York Public Library resulted in teens using iPads and smartphones to create a game. Using the online platform

of SCVNGR, participants created clues and riddles about their local neighborhood history. The Edgar Allan Poe Cottage was located in their neighborhood. The teens learned more about Poe and his legacy through this game (Martin, H. Jack, 2011).

- **Educational Game Technology**, South Carolina State Library. Many libraries lend mobile games and/or their devices to patrons to check out just like they do books with their library card. The South Carolina State Library is just one example of a library that lends out DS Lites and games to member libraries within the state. While these aren't lent out directly to patrons, libraries request the mobile games and devices which then can be used to host a program in the library for patrons to use (Hotchkiss, Deborah, 2007).
- **Library Conferences, Location Based Services**. Librarian Joe Murphy shares on his blog post how he's used location based services such as Foursquare, Gowalla, and Getglue at major library conferences to create experiences for attendees to learn together and obtain rewards for playing (Murphy, Joe, 2011).

QR codes are just the beginning of how public libraries are using mobile devices for gaming. Even though gaming consoles such as the Xbox or PS3 offer so much more than just gaming such as the ability to play movies, they still have a stigma attached to them in some libraries, particularly in the difficult economic picture of recent years, that gaming isn't a priority activity in terms of how we should be spending our time and scarce resources. However, integrating gaming with mobile devices such as phones or iPads, in libraries, might have a bit more credibility than with using a console because the mobile devices can be used for so much more. IPads, for example, could have apps that help deliver book related content in a different way. Phones can access apps that are a direct link to a library catalog or contacting a librarian. In other words, the relationship between gaming and accessing information via mobile devices might have a more obvious link to the priorities of a library such as literacy or educational success.

Because of how the economy has affected the viability of libraries in the last five years, to be innovative yet with less staff, and still deliver services that are essential to many communities, can be challenging. While video gaming has become a more mainstream activity in many public libraries, there's still the need to prove that this is an important activity where participants are learning valued information that can't necessarily be replicated in the same way by other organizations in the community. This isn't to say that there is any less learning that is valued by the library going on when engaging with console gaming, but that the learning is somehow more apparent when using mobile devices because of their multiple uses and also the ability to bring geography into the picture. Many games, including scavenger hunts using smartphones or GPS units are dependent on location.

For an organization such as a library that seeks to bring people together with the larger community while integrating such things as neighborhood geography whether through an actual location of a clue or simply anytime access to information, this can be a very powerful tool for libraries to help make these connections.

## references

Barack, Lauren (2012). To Attract Parents and Kids, Libraries Should Think Mobile. Retrieved from http://www.thedigitalshift.com/2012/01/research/to– attract– parents– and– kids– libraries– should– think– mobile/

Carlucci, Lisa. (2012). The State of Mobile in Libraries 2012. Retrieved from http://www. thedigitalshift.com/2012/02/mobile/the– state– of– mobile– in– libraries– 2012/

Danforth, Liz (2009). Mobile Gaming. Retrieved from http://blog.libraryjournal.com/games-gamersgaming/2009/09/10/mobile– gaming/

Find the Future at NYPL: The Game (2011). Retrieved from http://game.nypl.org/#/

Hotchkiss, Deborah (2007). South Carolina State Library Public Libraries Can Check Out Educational Game Technology. Retrieved from http://www.statelibrary.sc.gov/public-libraries– can– check– out– educational– game– technology

Karp, Josh (2010). The Chicago Public Library Helps Teens "Find History". Retrieved from http://spotlight.macfound.org/featured– stories/entry/chicago– public– library– helps– teens– find– history/

Martin, H. Jack (2011). The New York Public Library: NYC Haunts: Bronx Teens Discover Their Neighborhood Through an Interactive Look at the Dead. Retrieved from http://www.huffingtonpost.com/the– new– york– public– library/nyc– haunts– bronx– teens– di_b_898206.html

Murphy, Joe (2011). Foursquare & Conferences: Enhancing library conferences with Location– based services. Retrieved from http://joemurphylibraryfuture.com/foursquare– conferences/

Nelson, James. (2011). "Bubble Ball" iPhone app inventor is Utah eighth grader. Retrieved from http://www.reuters.com/article/2011/01/20/us– whizkid– iphone– idUS– TRE70J05W20110120

Tarr, Russel. (1998– 2012). How to Set up a QR Code Treasure Hunt. Retrieved from http://www.activehistory.co.uk/Miscellaneous/menus/history_mystery/qr.php

Topeka & Shawnee County Public Library– Topeka, KS Foursquare (?). Retrieved from https://foursquare.com/v/topeka– shawnee– county– public– library/4b06c23cf964a520b6ef22e3

mobile libraries

kelly czarnecki

build your own

# planning your game jam: game design as a gateway drug

*by Colleen Macklin, John Martin and Seann Dikkers*

# planning your game jam: game design as a gateway drug

*by Colleen Macklin, John Martin, and Seann Dikkers*

If you want to truly understand a topic, design a game for it. This chapter provides a framework for running your own 'game jam' — or, if you only have a small chunk of time — a game design challenge. A game jam is a game design tradition to gather a range of prospective game designers that work in groups to make a fully- functional game, from idea to iteration, within a set amount of time. The advantage of community- based design models like this one not only builds creative collaboration, but also allows for playtesting and natural enthusiasm around game design. It reminds all designers, no matter how accomplished, that an idea is only as good as its execution, and that productivity is measured in moments, not months.

## INTRODUCTION

A decade ago the first game jam took place in Oakland, California prior to the annual Game Developers Conference as a means to encourage experimental game design around a theme. The 0th Indie Game Jam's theme was "10,000 guys," or, how can one design a videogame with 10,000 computerized characters? The idea caught on. Since then, game jams have taken place at schools, game design studios, and basements around the world. Ten years later, in 2012, the weekend Global Game Jam included participants from 47 countries around the world. Over 10,000 designers created games themed by an image of the Ouroboros, the serpent eating its own tail. Locally we have used game jams to address civic challenges, explore digestive systems, or simply to have a good time and build relationships. We suggest that this basic approach can be used in a variety of ways.

First, the official breakdown of a game jam:

### Official Game Jam Rules:

*Officially at each site, the Global Game Jam runs continuously for 48 hours in each time zone, beginning at 5:00 PM on the start date, and ending at 5:00 PM two days later. The recommended schedule includes a short planning and team creation period, followed by development time until 3:00 PM on the final day. The last few hours are set aside for teams to present their creation to each other. However, sites are not required to follow this schedule.*

*At the beginning of the event participants are given a theme, such as "Extinction" in the 2011 Jam or "Ouroboros" in 2012. Participants are asked to create a game that in some way relates to this theme. Additionally, participants are given a list of "achievements," also referred to as diversifiers. These are designed to drive creative development by adding a unique or limiting factor to their game's design. Examples include "Both Hands Tied Behind My Back," in which a game should be designed to be played without the player's hands, or "Picasso Lives," in which game art must be cubist in style.*

Of course, any and all of these rules can be modified to fit the local game jam context and goals. Themes and achievements can both be used as ways to challenge jammers further, or to promote a learning agenda by the organizers and differentiate the challenges.

Collectively we have organized and facilitated jams of various kinds involving university students, teachers, researchers, hobbyists, storytellers, and even game designers. There's no design experience necessary and no technical expertise either. Jams can be done on paper or in computers. Really, all that's needed is a willingness to set aside some time, a space to meet, and let folks know. Customize the rules, themes, achievements, time, and other variables to make the event as stimulating as possible for participants — together we call these pedagogical tools *constraints*.

The framework in this chapter uses both technical and conceptual constraints. Constraints are the designer's best friend; they provide challenge and provocation, just as rules provide players with a framework to play and innovate within. The constraints in this framework are modular so that they can be reconfigured to provide new types of challenges and creative prompts.

In fact, the framework presented is modular enough to enable shorter or longer jams and the ability to string together a series of activities over the course of a couple of hours, a day, or a week. It's designed with the assumption that many might not want to stay up all night drinking Red Bull and debugging code (although this is rewarding, to some). Core activities include:

1.  **Ideation.** How to create "generative constraints," brainstorm, and identify one idea for your game. We have both had participants throw themes into a hat then draw two, and chosen idea prompts ahead of time and advertised them so participants can start to generate ideas ahead of time.

2.  **Prototyping.** How to build a paper prototype in the first few hours of the jam and use the remaining time iterating and refining your core concept. Paper prototyping is one of the most important techniques for all designers, even if the final design is paperless. In addition to paper prototyping, specific prototyping techniques for location– based mobile games will be described, from a walking prototype to using scaled– down versions of the game's location(s). In these cases available resources (string, sticky notes, markers, maps) also serve as a variables.

3. **Playtesting I.** The only way to truly understand the dynamics of your game is to playtest it. And the key to playtesting is early and often. Playtesting will help you/your team "find the fun" in your game and prioritize design decisions so that your project stays within the scope of something you can actually make and something people actually want to play.

4. **Prioritizing.** Figuring out what is the most important aspect of a game and building from that, while slashing everything else, is one of the most painful endeavors even for the most experienced designers. It's also a secret to the success and failure of most games. During a fast– paced game jam it's even more critical to prioritize. The activities in this section will help make it easier to get into production with a design that is actually doable!

5. **Production.** This is where all of the elements come together and the true test of the technology — and your skills — come into play. For the Locative Game Jam we'll narrow down the technology to choices that beginners will be able to work with and more advanced designers will be able to push further. Look for tools that multiple people can use at the same time like butcher paper or Google Docs.

6. **Playtesting II.** Playtesting your game while in production serves several functions: it helps you see whether the technology is working before you build too much, it helps you find bugs, and it keeps you honest. The proof of a game is not in its graphics, code, or concept, but in its play(test).

7. **Show and Tell.** At the end of the game jam, typically there is some kind of large group sharing of games made, features created, and a collective introspection of the process.

By the end of the jam, participants will have learned the design process from start to finish and will have made a fully– playable game. Whether the jam takes place over 24 hours or 24 days, it's a great introduction to the elements of game design and to thinking like a designer. Even a 1– 2 hour activity that focuses on prototyping, with some time for reflection, can be a valuable exercise in design thinking.

## ORGANIZING YOUR OWN GAME JAM

Game jams are best mastered through practice. We want to make that as clear as possible to encourage those of you interested in organizing one for yourself. You will learn the benefits, fun, and management of game jamming as you do them. With each one, you'll add ideas, style, and specifics that will make the next one you do even better. If you have access to a room, basic office supplies, and a group of willing, creative people, you are set to put together a game jam of your own. In the words of the Nike ad campaign — 'Just do it'.

At each stage, the following points provide an overview of planning considerations before, during, and after you run your own game jam.

Before your jam:

**1) Goals.** As noted above consider the goals of the jam. Whether you are motivated to build community, learn game design, or engage with content, your goals change the specifics. Goals affect time allocation, space, and materials that are needed below. For each point below use these goals to guide your decisions.

**2) Audience.** Audience also affects the planning of the jam, yet, in our experience, not that much. For younger audiences, part of the appeal of a game jam is that the workflow resembles adult work teams and project– based careers. The high energy of deadlines and creativity are a good fit for younger audiences. On the flip side, adult audiences are attracted to the low– stakes, playful, and also high energy environment.

The background and motivation of the participants are more important than age. Consider before you begin the reason for attendance on the part of participants. If you have willing volunteers that have previous jam experience, much less needs to be done up front to explain and contextualize the jam. If your attendees are there because of a class or other requirement, you'll need to plan for more explanation, clear goals, and think about motivation for participation. As in any creative space, unwilling group members can stutter or stagnate production. In compulsory settings, plan accordingly to allow for free participation in groups and an alternative activity in the rare case that someone wants out.

**3) Space.** Select and reserve space for the jam that best maximizes the process. Game jams are best done apart from day–to–day work. Consider moving off–site, or rearranging the site, if you have a group that works together already. For larger communities that may not already know each other, a quality space adds to the excitement and prestige. We've seen that large open spaces work well, especially when you can easily break off into smaller planning groups without losing line–of–sight of others. When one group starts using butcher paper, whiteboards, or index cards to organize thought, other groups should be able to look across the room and get ideas for collaboration and process.

Your space should facilitate central focus for initial kick–off presentations and projecting final game projects at the end. Chairs need to be generally tolerable or plan for stretch breaks. Check for available parking, clear directions for commuters, and provide signs inside the building with clear directions to the room. Game jams that take longer periods of time will need easy access to food, restrooms, and perhaps hotels. Finally, we encourage that food be allowed in the room for groups that want to work through meals together — wrappers can make fine decorations.

**4) Materials.** Our game jams have ranged from entirely paper and marker to full digital builds of games. Using your goals, audience, and space consider what materials best meet your goals of design time. For a non–digital jam, make sure you provide a ready supply of sticky notes, markers, string (to show links between game elements), tape, and paper. In addition, groups that have whiteboards or flip chart paper can easily share ideas and keep plans 'in front' of them. Consider other fun elements like board game pieces, stickers, and anything that can add a flair to either idea generation or planning.

If using digital tools, have provision for everyone to plug into an outlet and access a wireless connection. Be wary of the capacity of the wireless network. Find a way to test it, contact IT, or find another space where you can confirm the network will handle the work flow. We have found it best to avoid computer labs because, as a space, they generally are set up for individual work, not group work. Of course, if you are providing laptops or mobile devices for the jam, make sure these are fully charged and access to charging stations is available.

During your jam:

**5) Opening ceremonies.** For veterans or newbies, game jams should start with opening comments from the host. As you prepare your notes, include appropriate "Thank yous," welcomes, housekeeping, and rules for the facility use. Point out where participants can find food, restrooms, and any needs once the jam starts. The guts of the opening comments should include the parameters of the game jam or the design constraints. Include the times and overall agenda for the session, what the goals of the jam are, and the overall constraints placed before the participants. You can also use this time to organize groups, but we recommend having groups established ahead of time. Your opening comments should end with, "...and GO!"

**6) Brainstorming.** Plan for groups to meet and begin brainstorming around the design constraints of the jam as soon as possible. Allow roughly 20% of your time for this process and use any number of available strategies for keeping it fresh and pushing generativity beyond the initial rush of ideas. Many game jams have a period of large group sharing of ideas from each group to help narrow the ideas that they are excited about and compare, adopt, and redesign based on hearing ideas from others.

**7) Prototyping.** After groups settle on an idea, they should begin a process of fleshing out the game idea into specific stages of play. Allow 20%– 40% of the time for prototyping. In shorter or non– digital jams, a well worked paper prototype may even be the final product. A few ideas for prototyping include storyboards or 'maps' of the game, level concepts, 'wire– frame' drawings of screens the player would see, scripts, slide shows (with buttons!), or index cards organized to show game options. Prototyping can be non– digital or digital. By visiting the jam groups, organizers should be able to quickly get a feel for the game that is to be made, where the ideas have challenges, and how the group is solving problems. Be prepared to clarify and answer questions, encourage teams, and drop in additive ideas as you visit the groups. For younger jams, you will need to address group dynamics more, help them improve collaboration skills, and resolve conflicts.

**8) Build the game.** In larger game jams, building the game is the core of what motivates participants — taking up to 50– 70% of the scheduled time. For mobile media learning, building the game requires programming skills or the use of a rapid prototyping tool (like ARIS) to build working versions of the game.

**9) Test the game.** The benefit of having multiple groups designing together, is that you can stop everything, bring the large group together, and play each other's games. This may only take 3– 5% of the time. Having the same constraints, groups can learn from each other's ideas, solutions, and resources accessed. Develop a process or direction for playing games by either assigning a rotation, picking groups, or any method designed to have jammers playing other games. We suggest having the groups split between "stay with your project and present it" and "rotate to the other projects" so each group has a mix of interactions.

**10) Refine the game.** After building and seeing other teams' games presented, groups need a chance to return to their own game and apply lessons, ideas, and suggestions to their own game design. This time should be roughly a third of the time they took to build the game or 10– 20% of the jam time. This is also time to add final polish and prepare to present the final product to the large group.

**11) Closing ceremonies.** Bring the full jam group together for final presentation of the games. Plan for a central location for others to access and play games (e.g. a website, social document, or public display space), and time to get feedback for future jamming. We've tried and largely abandoned awards, finding that they don't necessarily fit with the motivation and goals for coming to a game jam — namely community, production, and fun. Yet, depending on the size of the groups, highlighting each game is a fun way to tie things together.

## POST–JAM:

**12) Follow up.** You learn how to organize jams by doing them. Take time after the jam to contact participants for suggestions for the next one. Because they are fairly informal, most participants are more than willing to help you design or even help plan the next one. Those with the most comments may be unwittingly volunteering to lead the planning committee.

**13) Audience.** Find a way to highlight or make game jam products accessible (with permission) to a larger audience. If you have an online site, this may mean populating it with links to games. If you are working with school groups, the local library may be willing to set up a kiosk or endcap with your games. Building an audience and community is important; more people means more relevant games, but it also increases connections between designers and encourages new participation.

## BUT I DON'T HAVE A WEEKEND!

"This is all well and good," you say, "for people with weekends to burn. But what can I do within a classroom schedule?"

It's a fair question. And it's one that we've been thinking about and experimenting with for years. Remember what we wrote about constraints, "Constraints are the designer's best friend; they provide challenge and provocation, just as rules provide players with a framework to play and innovate within." We hope that in reading the final section of this chapter you are encouraged that:

1. Great learning can happen through prototyping;
2. You can prototype a game in a couple hours (see lesson plan below);

Then we hope that in running a game jam you we see for yourself that:

3. Learner reflection during a game jam is potent and worthwhile, and
4. Learner reflection after a game jam has remarkable staying– power.

### a shorter 'lesson plan'

Let's be clear — this isn't as cool or immersive as participating in a Global Game jam for a weekend; nor is it even as good as a 3– 4 hour design jam. If you can do either of those, don't even bother with this! But if you're like many educators we've encountered over the years, who are interested in having their students design games, but have to operate within a couple school days or a school week, here's a game jam plan that has worked fairly well:

### students will be able to (SWBAT)

Work together to design and present a systemic understanding of the content through gaming media mechanics that models the topic.

## supplies

Pencils, colored markers, 3x5 index cards, scissors, assorted dice, 11x17 paper (game boards), small sticky notes, assorted playing pieces and tokens such as glass beads, stones, and poker chips. (The dollar store is your friend here.)

## Introduction (15 minutes)

- Briefly explain the activity: each group of ~4 will design a game for the others to play at the end of the hour.
- Hand out small scraps of paper and ask them to write down potential game themes such as "Dinosaur Ninjas" (we found that if you don't seed semi–ridiculous themes, students tend to submit recent course themes, which tend to make the challenge too complicated — see above). Pick one at random to be the theme for the games. (These randomized constraints help even the field of game ideas by eliminating themes that participants might arrive with.)
- Break into groups of about four people.

## Ideation (15 minutes)

Come up with many ideas, and choose a "doable" one — not necessarily the best one. As they come up with ideas, groups should consider the following:

- Guiding metaphor: Components. Break it down.
- Game play: What will it look like when they're playing?
- Prep: What has to happen before anyone starts playing?
- Game board: Where is it played? What are the physical constraints of play?
- Rules: What are the rules of play? Keep it simple!
- Winning: How does one win? Or, When does it end?

## Prototype 1 (15 minutes)

- Stop planning and make something!
- Try it! — Play along; it won't be great yet, but give it your best optimistic try.

- Is it fun? What could be better?
- Adjust it, or try something else, but remember that in this constrained time frame, a single "okay" idea is better than a pile of discarded ones. One strategy that we've seen work surprisingly well is when groups embrace what they might think is a bad idea, and pump up the ridiculous parts of it to make it so bad that it's fun! A mediocre game mechanic can be amped up with other constraints (e.g. "player must be blindfolded" or "all in– game communication must be sung in operatic style" etc.)

## Prototype 2 (15 minutes)

- If the first prototype was really terrible, here's your chance to take what you learned and go in a new direction with it.
- If the first prototype was workable, here's your chance to supplement it with a additional factors that enhance it.

If you have to spread this over two periods or sessions, here's a good place to stop. If participants are into it, the next step "Finishing Touches" (or even a third prototype) sometimes happens before the next meeting. If not, give the designers a few minutes at the start of the next session to refamiliarize themselves with their work, and add some finishing touches to it.

## Finishing Touches (15 minutes)

- Give your game a title if you haven't already. How does the title affect the game play, or pieces, or general look and feel of the game? Tie it all together!
- Give the latest version another quick playtest. Look for dead ends or actions that break the game and figure out how to roll them in to continued gameplay.
- Write down the rules. As much as possible, the rules should be as intuitive or implicit as possible, with clues on how to play built into the title ("Ninja Dinosaur Teradactyl Toss!"), or theme, or gamespace — but any additional rules should be clearly written.

### Group Play! (30 minutes)

There's a lot of room for customization here. You should allow at least 10 minutes for each game. That said, we've seen some great 2– player games that one can get a good feel for in a few minutes, and some great 12– player games that participants want to play completely through. The underlying goal is to provide enough time to get enough of a sense of the game that one is able to provide feedback to the designers in the "Reflection" section.

### Reflection (15 minutes)

In many ways, this is the most important part. Provide some time for groups to give each other feedback on the game, both individually, and as a group. Discuss what worked, and what didn't, and what might work better. This is where designers learn from each other and where you get to bring home more complex points and issues that you noticed during the process.

Before you take off and start planning, we have a couple closing thoughts to share. First, we hope you can sense the enthusiasm that we have for this process. Group work isn't new at all, but game design and the process for game jamming has taken off because it is a fresh way to enjoy a weekend, after–school program, or even a classroom activity.

Game jam planning is relatively easy and the payoff has exceeded our expectations. However, as with any new approach to learning, you will most likely run into snags and challenges. We haven't spent as much time discussing challenges because we just haven't run into many, and because the mobile project chapters cover many of them well.

Finally, if you are setting up a game jam and want to reach out for someone to talk it through, look at your agenda, or just to share your experience, we are available. Look us up online and send us a quick note. We want to emphasize that we see mobile media learning as an emergent and growing community of people innovating — we'd love to meet you!

In the meantime, enjoy your game jam.

youth applab:
the wonder of app inventor
and young app developers

*by Leshell Hatley*

# youth applab:
# the wonder of app inventor
# and young app developers

**by Leshell Hatley**
*Uplift, Inc.*

Youth APPLab won the MacArthur Foundation's 2010 Digital Media and Learning Competition and set out to expose African– American students in the District of Columbia to cutting– edge technology, empowering them to design and develop mobile applications (apps) as opposed to only becoming consumers of them. The first and longest running after– school program of its kind, Youth APPLab presents participants with opportunities to explore what it means to be a computer scientist, to examine various forms of cutting– edge technology, to innovate, problem solve, and express themselves. Occupying a unique space in the collection of after– school programs, Youth APPLab blends computer programming instruction with research– based inquiry into student perspectives of software engineering.

The first year of Youth APPLab launched with 22 students and a long enthusiastic waiting list. Although eager to begin, approximately 90% of this initial cohort had no programming experience and almost all did not know what it meant to 'code' (a word commonly used to mean the writing of computer programs). Perhaps just one of the students in this cohort had a smartphone and, although many had computers at home, none had ever conceived of creating a mobile app themselves, until hearing of Youth APPLab. At a time when computing technology impacts almost every aspect of our daily lives, this unfortunate lack of exposure to the creative power of technology permeates through many communities of color and is at the heart of why Youth APPLab

FIGURE 1. YOUTH APPLAB INSTRUCTORS AND STUDENTS

was formed (See figure 1. above).

Besides being an amazing testament to students' hard work, creativity, and enormous accomplishments, this chapter shares with the reader valuable suggestions, best practices, and some of the key ingredients essential to the success of our work. In the following pages, I'll provide an overview of Youth APPLab, our curriculum, and how App Inventor played a major role in the implementation of the first year of this after– school program.

## WHY YOUTH APPLAB?

We hypothesized that by providing a supportive environment, enriched with culturally relevant pedagogy, high expectation, and opportunities to learn from peers, students of color would become engaged, highly motivated and persistent, and interested in learning more about computer science and becoming computer scientists. We imagined that exposure to computer programming instruction for almost a year and an opportunity to intern the following summer would ignite a passion to learn more about the field. After 10 months of intense instruction with 1 elementary, 8 middle, and 11 high school students (7 girls and 13 boys), our hypothesis was proven and the results were beyond our richest imagination!

## KEY INGREDIENTS FOR SUCCESS

Hosted in one of Howard University's computer labs and armed with a seasoned STEM (science, technology, engineering, and mathematics) instructor, a professional software developer who served as co– instructor, and several guest speakers, Youth APPLab's curriculum was delivered with high expectation, important culturally relevant messages and examples, a sprinkle of entrepreneurial concepts and encouragement, and tons of insight from many Howard students currently majoring in computer science. We later realized that Youth APPLab students made specific connections with these college students as they shared a great deal in common (e.g. African– American, high school, the desire to major in computer science). As such, they listened intently to their insight, suggestions, and instructions.

These assets and resources proved extremely beneficial to students, helping them form positive attitudes about computer science overall and contributed greatly to their development of positive dispositions about computer programming, problem– solving, and troubleshooting. These combined with an awesome tool for beginning programmers, called App Inventor, greatly increased student efficacy towards creating their own mobile applications.

mobile software

leshell hatley

## APP INVENTOR

App Inventor is a web– based visual programming language used to create mobile apps intended to run on Android devices. It was created collaboratively by Google and MIT and is freely available to the public. App Inventor offers its developers the use of drag and drop elements that are easily moved and manipulated on screen. These drag and drop elements replace the traditional text– based programming languages that have rigid syntax rules, such as Java (the traditional programming language used for developing Android apps). It is commonly known that text– based programming languages can be extremely complicated and frustrating for beginning programmers to learn. Therefore, the simplification offered by App Inventor makes it an optimal, concise, and effective entry– level development tool for any beginning programmer. As such, it fit nicely into Youth APPLab's curriculum.

App Inventor is made up of two software modules that work together to aid programmers in mobile app development. The first is a web– based graphical user interface (GUI) called the 'Screen Editor,' [Figure 1, p.211] used to:

1.  design and layout the interactive features (components and properties) of app (e.g. images, buttons, titles), and
2.  assign the use of physical mobile device features (e.g. sensors and screen orientation).

The second module runs directly on the programmer's computer and is called the 'Blocks Editor' ( see Figure 2). It is used to drag, drop, and manipulate puzzle pieces called blocks, each with their own assigned color, shape, and programming logic. Blocks are intended to fit together with other blocks in order to create portions of a mobile app's program. These program portions bring functionality to the manipulated device and graphical features designed using the first module, the screen editor. Forming one complete program, these blocks of code are compiled (tested for errors) and are made available for immediate download to a physical smartphone or to App Inventor's device

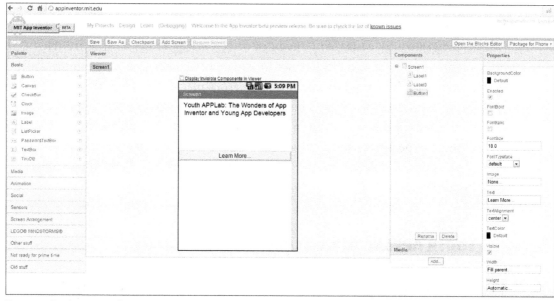

FIGURE 1: APP INVENTOR: SCREEN EDITOR

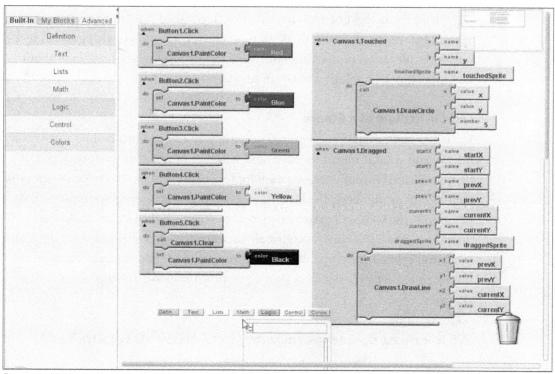

FIGURE 2: APP INVENTOR — BLOCKS EDITOR

emulator for testing. Although designed to make teaching and learning how to program a bit easier, App Inventor does not negate the need to troubleshoot — even though blocks only fit with other blocks that programmatically go together. The need to troubleshoot errors may still exist if the app's logic is designed incorrectly or a design element is unexpectedly incorrectly connected. In this vein, App Inventor still provides authentic software development experiences and produces high quality apps. So much so, that some professional developers and hobbyists sometimes prefer to use App Inventor instead of Java.

## YOUTH APPLab IMPLEMENTATION

Youth APPLab students attended 2– hour classes after– school, two days a week from October 2010 to June 2011. Students were equipped with laptops running Ubuntu Linux to provide an enriched computing experience on a regular basis, as most of the students had never seen the Linux operating system before Youth APPLab. Beginning classes were filled with general technology exploration topics. Students were encouraged to ask questions about any aspect of technology they desired.

### Our Curriculum at a Glance

Initially, class discussions centered on piracy and why downloading music was illegal, and what it meant 'to hack' and whether or not that was good, bad, or spy– like. Eventually lessons explored software and hardware concepts as well as binary numbers, the history of computers, operating systems, desktop programs, the web, and eventually grew to what Youth APPLab calls the 'mobile– verse.' The conversations about software grew into software development and related concepts like algorithms, pseudo– code, human– computer interaction, user experience design, agile programming, testing, and troubleshooting.

While learning these terms, students learned Alice, a 3D visually– based programming language designed for beginning computer science majors made at Carnegie Mellon University, and Scratch, the first block– based visual programming language made at MIT. Links to all resources listed here can be found at the end of the chapter.

After a few months, (with parent permission) students were given fully functional Android smartphones, with 100% voice, text, and data 24– hours a day. They were immersed in the 'mobile– verse' and were tasked with exploring the Android Market (now called Google Play). This provided students with insight into what it means to have and use a smartphone and have access to its unique features (GPS, accelerometer, etc.) on a regular basis. Students were asked to install and uninstall apps at will, making note of what was appealing about those apps as well as what was not so appealing. These notes provided great insight and would later be referenced when students designed their own apps. A website was created where students blogged after each class period and also posted official app reviews on a weekly basis. Blogging and writing official 'product' reviews are a noteworthy replication in a variety of educational settings. They specifically prompt focus on writing, self– expression, and other essential communication skills. However, since Youth APPLab is not primarily focused on literacy instruction, we guided these exercises with short, formal instructions and several examples for students to model.

## APP DESIGN & DEVELOPMENT

App design began after two months of smartphone immersion and instruction in the fundamentals of computer programming. Students were encouraged to be creative and to design and develop whatever appropriate app ideas came to mind. To do this, students learned the development lifecycle. Although, there are several approaches that can be used as a software development project progresses through the various stages needed to bring an idea to fruition, the size, scope, skill, and preferences of the development team usually determine which approach is used. We taught Youth APPLab students a general series of development stages: brainstorm, design, develop, and test. Often our youth designers needed multiple rounds through stages in order to reach an error–free program. As they progressed through each stage, students also created various paper– based and other resources needed to help them progress successfully. These resources included age— and experience— appropriate design documents and other templates as well as collaborative methods used by those students working in teams.

An app design method I created called 'Moving Panels,' was used to provide an easy transition from traditional paper– based designs (i.e. drawings on paper with colored pencils) to conceptualization of each aspect and interactive element of a mobile app. Moving panels basically consists of a notepad with several blank pages taped to a block of wood covered in aluminum foil. The block of wood simulates a mobile device in size and feel in the palm of one's hand. Each page of the notepad features the various screen changes that occur when running a functioning app. After learning how to use App Inventor,

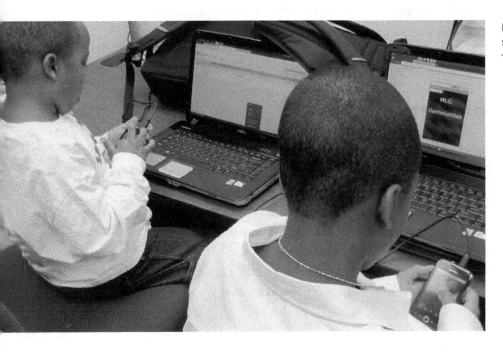

FIGURE 4. YOUTH APPLAB STUDENTS CREATING AND TESTING APPS.

FIGURE 5. YOUTH APPLAB
STUDENT ALLOWING OTHERS
TO DEMO HER APP.

students transferred these screen designs directly and easily into the screen editor in App Inventor.

Due to student interest and some scheduling conflicts at various times throughout the year, a third class day was added each week used as non–instruction open– lab days, giving students dedicated class periods to work towards completing their apps. Students worked vigorously towards the completion of their apps until the last day of class in June 2011, cycling through rounds of development and testing (see Figure 4). The school year culminated with a community 'Demo Day,' where students shared their Youth APPLab experiences and demonstrated their apps with enthusiastic and supportive family members, friends, and other invited guests.

Demo Day was extremely important to our students and their families. More-over, it also provided a culminating and authentic assessment experience for students. Tasking students with the creation of real– world mobile apps and the steps taken to develop, test, and publish them, were extremely valuable to student learning and to our intended program outcomes.

## RESULTS

By Demo Day, Youth APPLab students created more than 30 apps, 4 of which were published on Google Play (the Google Android Market) using our Google Developer's account. Combined, these apps have a total download count of over 5,000. Many more apps will be published summer 2012.

Student creativity with app development was incredible and many of them emphasized some type of information sharing and instruction with intended app users. Several ideas even went above and beyond the functional features of App Inventor and when asked to scale these ideas down, many students opted to create additional ideas as opposed to downgrading an original one.

Resulting apps included several learning tools and tutorials (e.g. alphabets and numbers in 3 different languages with sound, 3rd grade math, human body parts, Egyptian deities, and an App Inventor tutorial app); reference apps (e.g. suggestions for how to get accepted into college, skateboard park, and book price locators); games and sports (e.g. apps called ColorTap, Robo– smash, and Sneaks on Feet); and more (e.g. apps to control robots). A link to all apps on the Google Play's app market can be found at the end of this chapter.

Two brothers focused on the entrepreneurial concepts taught in class and surprisingly formed their own mobile app development company and with it, now have their own apps published on Google Play as well. These accomplishments and the confidence, pride, and foundation in computer programming that come along with them are a testament to the power and ease of use of App Inventor. Students' ability to conceive of an app and bring it from an idea to an actual app running on an Android device became second nature and many have made several apps outside of Youth APPLab since then.

One specific experience that showcases what students learned about app design and development using App Inventor occurred when two students attended the 2nd Annual Digital Media & Learning Conference in March 2011 to share and demonstrate Youth APPLab and their experiences to– date. As part of our demonstration, these two students, Hamza Hawkins and Kweku Sumbry, requested to receive app ideas via twitter and challenged themselves to make and test the app before the 3– day conference was over. Selecting the target app from Youth APPLab's twitter timeline (@youthapplab) on day two of the conference, Hamza and Kweku decided to create a suggested app that would send business card information and append a note. Hamza and Kweku's experience with App Inventor to– date was so ingrained, that they finished and demonstrated the app to 100+ conference participants in less than 30 minutes. This app is undergoing a design upgrade and will be made available by the summer 2012.

Some students joined Uplift, Inc., the nonprofit organization which powers Youth APPLab, as summer interns after Demo Day. They applied all they had learned working collaboratively as a youth–based mobile application development company. Twelve students worked in this capacity for 7 weeks, working Monday through Friday, from 9am– 5pm, and were compensated to make 5 additional apps. One of these apps was presented and discussed at the 3rd Annual Digital Media & Learning Conference in March 2012.

The first year of Youth APPLab was so influential that all four graduating African– American male participants entered college after having switched their focus to either major or minor in computer science. Additionally, most, if not all, of the other students experienced an increase in computer science and a desire to advance to the next skill level (i.e. many students specifically asked, "When can we learn how to program like this?" while wiggling their fingers as if to simulate typing on a keyboard). This showed that students were ready and willing to learn a text– based programming language and prompted the creation of more advanced programming classes. We found that Youth APPLab students were using App Inventor for assignments outside of Youth APPLab and, as a result, won competitions and even completed apps for their parents. All of the Youth APPLab students decided to focus on computer science or some aspect of it as a potential career. Uplift, Inc. was able to publish these and other additional findings in the first published paper about Youth APPLab.

It goes without saying that the instructors, students, and parents were extremely pleased with the outcome of this first year, enthusiastic support reached far and wide outside the walls of Youth APPLab and Uplift. We enjoyed a visit from FCC Chairman Julius Genachowski, a profile on Black Enterprise Television and other press. These external sources of support and celebration provided additional validation of the potential of each Youth APPLab student.

## LESSONS LEARNED

Youth APPLab has been a major success since initial launch and we are extremely proud of the road we're paving. As we look back at how far we've come, there are a few distinctive aspects about the project that we can share.

### be/have a Techie

First, although App Inventor is designed to introduce computer programming concepts to beginners (there are beginner tutorials online in various places), the instructor still needs to have a deep understanding of these concepts, should have created all app assignments before giving them to students, and should remain open about facilitating the learning and use of App Inventor. These characteristics will help tremendously when it comes to explaining computer programming concepts (e.g. algorithms and how to translate them into code), troubleshooting errors, solving overall problems, and allowing room for the imagination and excitement of budding programmers.

### go mobile

Second, although App Inventor comes with a software– based Android device emulator, we've found that seeing, testing, and working with an app on an actual mobile device is extremely rewarding to students — considerably more rewarding than when performing similar tasks on the emulator. Consider funding sources, storage, and maintenance of mobile devices as worth the investment if you choose to pursue a program like ours.

## scalability

Third, at times a project may involve the teaching and learning of MIT's Scratch visual– based programming language as first steps toward learning how to program, similar to the path we took in Youth APPLab. This is beneficial as Scratch and App Inventor are closely related and the skills and behaviors learned in Scratch are easily transferable to App Inventor. You can find links to these resources at the end of this chapter.

## make it 'real'

Lastly, as it relates to student work ethic, we noticed that students work their hardest when they are expected to work on a project that is real/serious (e.g. a context of social justice), has meaning, and has some level of difficulty. This, along with the reminder that final apps could be published for public download, put students in the mindset of working towards and persisting through to a worthwhile goal. As such, students will naturally enjoy the design and problem– solving process and will want to perfect and advance their skills almost without asking. This leaves a ton of budding software engineers out there just waiting to be introduced to App Inventor. Once you create a similar program to make this happens, we believe they will surprise you beyond your wildest imagination!

## how to contact us

If you are interested in creating a similar program or simply have questions about our work, please feel free to send an email to *bookresponses@youthapplab.com*.

## resources and their URLs

App Inventor — appinventor.mit.edu

Youth APPLab — www.youthapplab.com

Uplift, Inc. — www.upliftdc.org

Scratch — scratch.mit.edu

Alice — www.alice.org

Open Blocks — http://education.mit.edu/openblocks

Link to all Youth APPLab apps on Google Play — https://play.google.com/store/search?q=youth+applab

# stay in touch

## Editors

We believe that Mobile Media Learning is still in its infancy. The examples included here are only the beginning of a learning ecology that includes adventures, activism, collaboration, and moments of inspiration fostered by mobile, personal, collective, and increasingly affordable tools.

As a reader of this book, you are on the cutting edge of design for learning using mobile technologies — you now know what we know. You have the tools that we have. Our best advice is now yours to use as a starting point.

That means two things —

First, we consider you our colleague and we would love to know what you are up to. Please stay in touch. If you try a design or a jam, contact any of the editors or authors and share your experiences with us. (Just run a search online for any of us). We aim to collectively gather stories, accounts, and lessons so that we can more readily think of learning as mobile.

Second, we gladly turned down three other publishing houses in order to offer you the opportunity to add to this book. ETC press allows us to add chapters and do versioning of the book with relative ease. If you have an experience that adds to the scope of this work, and are willing to invest the time to write and revise, we would like to add your chapter to the book for future buyers to enjoy. Use chapter two as a guide for writing and shaping your chapter and send it to Seann.

Our next planned project is a follow–up to this work and your stories that attempts to organize and draw common planning and design principles across cases. Look for Mobile Media Design in the coming years.

For now though, probably have some mobile designing to do, we'll leave you to it.

All the best,

The Editors

# acknowledgements

Clearly, this book is a team effort. We couldn't have even dreamed of a project like this without a robust and collegial community of practice. First, we want to thank Kurt Squire for his suggestion to gather together these projects in one place. Thank you Kurt, for your guidance and encouragement through it.

This book wouldn't have happened without the amazing community of researchers, designers, and educators that have helped to build the Games + Learning + Society group. It's thanks to the leadership of Jim Gee, Constance Steinkuehler, Rich Halverson, Erica Halverson, and Kurt that we were able to easily meet, dine with, share, collaborate, and build many of the ideas found in this book. This work is the fruit of all the institutions listed with the authors and their participation in GLS - especially UW-Madison and MIT that have actively supported and championed spaces to do this work.

We are deeply indebted also to Drew Davidson and Scott Chen at ETC Press. We believe the work they are doing there to move a book from conception to print smoothly and quickly is setting a model for academic publishing. Books like this need to be timely, and ETC has made that happen for us. They have supported and piled hours of time on getting this text ready and all of the layout and style is theirs. Thanks.

Thanks also go to our authors. They have busy lives building and making amazing learning happen and taking a bit of time to write and rewrite is most appreciated. It is their work and ingenuity that make this book exceptional. We look forward to many efforts together in the future.

Also, this work wouldn't happen without a passionate, active, and forward-thinking players - students, educators, designers, and researchers. Mobile games would just be a dream without those that enjoy playing them, showing up to events, and encouraging us to make more. Thanks to all of you playing.

Finally, any project like this requires supportive family and friends. We most appreciate our support, comfort, and constructive critics that put up with late nights, red pens, and continual re-reading. You are all the best, and we love you.  Thanks.

# participant bios

## SEANN DIKKERS

*sdikkers@gmail.com*

Seann Dikkers is a researcher and recent doctorate graduate of educational technologies at the University of Wisconsin – Madison. Prior to his doctoral studies, he spent twelve years as a teacher and principal. Now he's serving as designer and consultant in new media education strategies for leadership and learning. Dikkers edited the recent release of Real-Time Research: Improvisational Game Scholarship. His work focuses on 21st century skills and tools, digital game engagement mechanics, educator professional development, and educational leadership. His projects include: ParkQuest; History in our Hands; the Mobile Media Learning project and Augmented Reality and Interacitve Storytelling editor (ARIS) with Kurt Squire; the Comprehensive Assessment for Leadership in Learning (CALL) with Rich Halverson & Carolyn Kelley; a game based history curriculum ('The American Idea'); consulting on digital tools for teachers; managing Gamingmatter.com; and raising two pretty awesome kids with his wife Stephanie.

## JOHN MARTIN

John Martin's heart is in expeditionary learning, and his doctoral research under Kurt Squire considered the use of mobile devices to connect people to the land and to each other at a deep woods camp in Maine. He now works for UW-Madison's Division of Academic Technology, and uses tools (like ARIS) and processes (like Digital Storytelling and game design) to support informal and formal learning environments and communities. He thinks people learn more by doing things than by studying them, and is excited that modern mobile devices have become Swiss Army Tools for learning and research. He blogs at times at *regardingjohn.com*.

**BOB COULTER**

Bob Coulter is director of the Litzsinger Road Ecology Center, a suburban St. Louis nature center managed by the Missouri Botanical Garden. He also serves as Principal Investigator and Project Director for two NSF-funded research projects conducted with MIT that leverage mobile technology to engage students in their local community. A key focus for his work is investigating the boundary between the virtual and the real, exploring ways to use technology to enhance students' understanding of the world. He is also writing a book on teacher agency, articulating a model of how teachers who lead rich community-based projects approach their work. In an earlier life he was an award-winning elementary math and science teacher, and he continues to promote recreational math through weekly volunteer stints in five St Louis area schools.

www.ingramcontent.com/pod-product-compliance
Lightning Source LLC
Chambersburg PA
CBHW080403060326
40689CB00019B/4117